Telling Tales in Sport and Physical Activity

A Qualitative Journey

Andrew C. Sparkes, PhD
Exeter University

Human Kinetics

Library of Congress Cataloging-in-Publication Data

Sparkes, Andrew C.
    Telling tales in sport and physical activity : a qualitative journey / Andrew C. Sparkes.
        p. cm.
    Includes bibliographical references (p.) and index.
    ISBN 0-7360-3109-X
        1. Physical education and training--Research--Methodology. 2.
    Sports--Research--Methodology. 3. Narration (Rhetoric) 4. Storytelling. I. Title.

    GV362 .S67 2002
    613.7--dc21

                                                                    2002017334

ISBN: 0-7360-3109-X

Developmental Editor: Joanna Hatzopoulos Portman; **Assistant Editor:** Derek Campbell; **Copyeditor:** Karen L. Marker; **Proofreader:** Julie A. Marx; **Indexer:** Sharon Duffy; **Permission Manager:** Dalene Reeder; **Graphic Designer:** Nancy Rasmus; **Graphic Artists:** Denise Lowry and Angela K. Snyder; **Cover Designer:** Jack W. Davis; **Printer:** United Graphics

10   9   8   7   6   5   4   3   2   1

Human Kinetics
Web site: www.HumanKinetics.com

United States: Human Kinetics, P.O. Box 5076, Champaign, IL 61825-5076
800-747-4457
e-mail: humank@hkusa.com

Canada: Human Kinetics, 475 Devonshire Road Unit 100, Windsor, ON N8Y 2L5
800-465-7301 (in Canada only)
e-mail: orders@hkcanada.com

Europe: Human Kinetics, 107 Bradford Road, Stanningley,
Leeds LS28 6AT, United Kingdom
+44 (0) 113 255 5665
e-mail: hk@hkeurope.com

Australia: Human Kinetics, 57A Price Avenue, Lower Mitcham, South Australia 5062
08 8277 1555
e-mail: liahka@senet.com.au

New Zealand: Human Kinetics, P.O. Box 105-231, Auckland Central
09-523-3462
e-mail: hkp@ihug.co.nz

To my Mum and Dad,
with gratitude, respect, and admiration—
but most of all, with lots and lots of love.

Contents

	Preface	vii
	Acknowledgments	xiii
	Credits	xiv
1	Surveying the Landscape	1
2	Scientific Tales	27
3	Realist Tales	39
4	Confessional Tales	57
5	Autoethnography	73
6	Poetic Representations	107
7	Ethnodrama	127
8	Fictional Representations	149
9	Different Tales and Judgment Calls	191
	Epilogue	225
	References	235
	Index	246
	About the Author	249

Preface

In recent years, as part of my personal journey as an enthusiastic but self-taught qualitative researcher, I have been thinking about the implications of regarding research as the receiving and telling of stories. Excited by the experimental writing that has been undertaken by scholars in the social sciences, where the findings from qualitative studies have been represented in new ways, I have also been thinking about the different kinds of stories that qualitative researchers might tell about sport and physical activity. This book attempts to pull together some of the strands of my thinking about traditional and new writing practices in these domains. As such, it needs to be taken as a "work in progress" rather than as a fully formed and polished end-product. My hope is that a book devoted to new writing practices will provide yet another voice in the fruitful dialogue on qualitative inquiry that is currently taking place in sport and physical activity.

Of course, my personal journey has not taken place in isolation. It has been framed by major changes in the social sciences instigated by postmodern critiques, the "narrative turn," and the dual crises of representation and legitimation that these developments have spawned. According to Bochner, narrative inquiry is both a turn away from as well as a turn toward:

> The narrative turn moves from a singular, monolithic conception of social science toward a pluralism that promotes multiple forms of representation and research; away from facts and toward meanings; away from master narratives and toward local stories; away from idolizing categorical thought and abstracted theory and toward embracing the values of irony, emotionality, and activism; away from assuming the stance of disinterested spectator and toward assuming the posture of a feeling, embodied, and vulnerable observer; away from writing essays and toward telling stories.
>
> *Bochner, 2001, pp. 134-135*

To provide a sense of these changes, I have divided this book into nine chapters that you can read independently should you wish to get a flavor of any of the particular tales that I focus on. The chapters have been organized, however, so that they work better when read in combination. Some of the tales will be familiar. Where this is the case, my tactic is to make the familiar strange by bringing to the fore the rhetorical features of such tales and then to illustrate how they operate to persuade you that they are worthy of attention. That is, I highlight how they are artfully constructed. With less familiar tales, I focus more on the reasons why various scholars have chosen to use these specific genres as a way of representing their findings, and also on how they have used each to good effect to generate different ways of knowing. I chose this tactic because I want to introduce you to the potential that different tales have to contribute to our understanding in sport and physical activity. I also hope the discussions provided in each chapter will assist you to make informed choices about why, when, and how you might engage with new writing practices in the future should you wish to expand your representational repertoire.

In **chapter 1,** to provide a backdrop to the emergence of new forms of representation, I focus on the shifting landscape of qualitative research in the social sciences in recent years. Key moments are identified before consideration is given to the ongoing crisis of representation and the attention this has focused on how research findings are presented in terms of their rhetorical and artful construction. Associated issues of reflexivity, voice, and ethics are also discussed in this chapter.

Chapter 2 focuses on the tale that dominates *quantitative* research into sport and physical activity: the *scientific* tale. Attention is given to the conventional models that structure and guide the writing of this kind of tale, and its historical roots are considered. The specific rhetorical devices and persuasive strategies that the scientific tale uses are then highlighted. In contrast, **chapter 3** is devoted to the tale that dominates *qualitative* research into sport and physical activity: the *realist* tale. Some unsettling similarities between the scientific tale and the realist tale are identified, and some suggestions are provided as to how the latter might be modified.

Chapter 4 deals with *confessional* tales. These foreground the researcher's voice and concerns about what happens during the actual fieldwork in a way that takes us behind the scenes of the "cleaned up" methodological discussions so often provided in realist tales. The personalized style and self-absorbed mandates of the confessional tale are carried forward into **chapter 5,** which focuses on *autoethnographies*. These highly personalized texts rely on systematic sociological introspection and emotional recall to allow authors to tell stories about their own lived experiences, relating the personal to the cultural. The charge of self-indulgence, which is so often leveled against autoethnography, is dealt with and refuted in this chapter.

Chapter 6 moves us into the world of *poetic representations*. Here, scholars write up interviews as poems honoring the speakers' pauses, repetitions, alliterations, narrative strategies, rhythms, and so on. This activity has been successful for many qualitative researchers in terms of helping them not only to rethink their data and analyze it differently but also to highlight their understanding differently.

The idea of *ethnodrama* is to transform data into theatrical scripts and performance pieces. **Chapter 7** focuses on this genre and looks at its strengths in terms of its ability to capture lived experience, reach wider audiences, and remain "truer" to life. *Fictional representations* are the focus of **chapter 8**. Here, scholars have chosen to express their understanding of phenomena in the form of stories. In *ethnographic fictions,* the notion of "being there" remains a central concern for making the claim that while the representation may be fictional, the data it is based on is "factual." In contrast, in *creative fictions* the authors give the narrative imagination free reign, and even though they may have "been there" in the field, they do not have to have been so, and they feel free to invent people, events, and places with a view to crafting an engaging, evocative, and informative story.

Leading scholars have raised questions of judgment regarding new writing practices in qualitative research. Accordingly, **chapter 9** foregrounds the ongoing crisis of legitimation and asks how we might begin to evaluate the different tales that researchers are now using to represent their work. The dangers of using inappropriate criteria to judge such work are highlighted, and more suitable ways of evaluating different tales, in the form of open-ended lists to be applied contextually, are discussed. Finally, the **epilogue** considers where the telling of different tales might take research into sport and physical activity in the future.

In each chapter, whenever possible, examples of work published in sport and physical activity are called on to highlight the analytical points being made about different kinds of tales. Given that new writing practices are only beginning to emerge in these domains, however, I have relied heavily on the work of scholars in other disciplines to provide frameworks for discussion and debate. At no place in the book do I advocate the elevation of one kind of tale over all others. My goal is to *displace,* not simplistically discard, classic forms of representation like scientific tales and realist tales, so that each becomes only one among many viable forms of social description. That is, I seek to *expand the narrative strategies* of qualitative researchers in these domains.

In seeking to expand the narrative strategies available to scholars, I do not intend to turn qualitative researchers in sport and physical activity into poets, novelists, or dramatists. Rather, following L. Richardson (2000), I intend to draw inspiration from these people to encourage researchers to acquire and nurture their own voices in their writing and to highlight writing as a process of *discovery, understanding,* and *analysis.* As Eisner comments in his reflections on qualitative research in the new millennium:

> There has been a growing realization in recent years among researchers of something that artists have long known in their bones; namely, that form matters, that content and form cannot be separated, that how one says something is part and parcel of what is said. . . . the form of representation one uses has something to do with the form of understanding one secures. Once this idea penetrated the research community, the form used to inquire and to express what one had learned was no minor consideration. This idea, the idea that different forms could convey different meanings, that form and content cannot be separated, has led to the exploration of new modes of research.
>
> *Eisner, 2001, pp. 138-139*

The astute reader will note that even though this book advocates experimental writing, it has been written in a standard fashion. Thus, it is a book about experimental writing rather than a book of experimental texts. I have chosen this tactic because, like Tierney and Lincoln (1997), I think the conceptual issues need to be laid out in sport and physical activity so that scholars in these fields can get on with the experiments themselves.

I am also aware that this book focuses only on written texts and has not investigated radically different representational practices, such as film, video, and other forms of communication that accompany new technologies. These have exciting possibilities and will receive greater attention in the future. Currently, however, the majority of scholars in sport and physical activity communicate their findings via written texts, so this seems an appropriate starting point to begin a dialogue about different forms of representation. As Plummer comments:

> Communication is at the heart of all social science: it is something which every social scientist has to do. From fledgling undergraduates writing a term paper or "doing a project" to doctoral candidates writing up their thesis; from conference paper to the formal scholarly journal article; from the "scientific text" to the myriad of writings that appear in books: every social scientist has to write! It is the *sine qua non* of social science.
>
> *Plummer, 2001, pp. 168-169*

Given that, as qualitative researchers, we do write, then we need to learn to write well. Unfortunately, in a review of the writing styles that inform a large number of ethnographic studies, P. Atkinson claims that atrociously dull writing characterizes much social science. For him, many texts are "by any criteria, inept and unreadable" (1990, p. 11). Indeed, talking of sociologists in general, Atkinson feels that many seem to revel in producing repellent texts that are distant, undecipherable, and dull.

The inability or unwillingness to communicate well is worrying, particularly given Wolcott's view (1994) that qualitative researchers need to be storytellers and that storytelling, rather than any disdain for number crunching, ought to be one of their distinguishing attributes. Related to this, Janesick notes, "For qualitative researchers, the story is paramount. And nothing is so important to the story as the words we use, both intuitively and creatively" (2001, p. 539). Likewise, L. Richardson comments, "Qualitative research has to be read, not scanned; its meaning is in the reading" (2000, p. 925). Therefore, if qualitative researchers don't tell a good story, readers will not engage with their work. At best, it might be scanned; at worst, it might not be read at all. These issues are serious, given the amount of effort qualitative researchers put into conducting their inquiries so that they can better understand the social world and then share this understanding with a range of different audiences.

Furthermore, writing is not just a habit for qualitative researchers but more a part of the definition of who we are. Therefore, we do not need or aspire to be literary giants to take pride in crafting our writing. This book is not designed to make you a better writer in any chosen genre. It is not a "how to" book about the ins and outs of writing. Excellent books already exist on this topic (for example, see Becker, 1986; Ely, Vinz, Downing & Anzul, 1997; L. Richardson, 1990; Wolcott, 1990; Woods, 1999). Rather, the intention of the book is to *describe,* not prescribe, how the worlds of sport and physical activity might be written about and known differently.

Improvements in any of the genres considered in the various chapters that make up this volume are a matter of hard work, commitment, practice, practice, and even more practice. As Wolcott (1990, 1995), one of the most talented writers of qualitative research in recent years, points out, he cares about his writing, he works diligently at it, he writes some and edits lots, and what others read are final drafts, not first ones. In short, as a skilled craftsperson, Wolcott, like the rest of us, has to work at writing.

Despite the hard work involved, I hope that this book will encourage qualitative researchers in sport and physical activity to experiment with how they represent their findings in the future, as part of an emerging research community that is spoken, written, performed, and experienced from many sites. I also hope this book will assist scholars in these domains to make reflexive, disciplined, principled, and strategic choices about when to use different forms of representation. Like P. Atkinson (1990), I will have succeeded if I have encouraged you to find a new complexity and a new source of fascination in your own writing as well as the writing of others.

Acknowledgments

In terms of the product, my thanks to Steve Pope for taking a chance with this book in the first place, and to Rainer Martens for taking on the risk. Thanks also to Joanna Hatzopoulos for her support as senior developmental editor. In terms of the journey, there are so many to thank, but here are a few to whom I wish to give a special thanks for sharing moments along the way: Pepe Dévis Dévis, Fiona Dowling, John Evans, Ken Fox, Deborah Gallagher, José Ignacio Barbero González, Phil Hodkinson, Mikko Innanen, Svein Kårhus, Herivelto Moreira, Jo Naess, Paul Schempp, Martti Silvennoinen, Brett Smith, John K. Smith, Mark Sudwell, Tom Templin, Artto Tiihonen, Richard Tinning, and Carmen Peiró Velert. My most special thanks goes to Kitty, my partner, and to my wonderful children, Jessica and Alexander, for the vibrant colors they paint on the landscape and for the joy they bring to my world. I love you dearly.

Credits

Angrosino text extracts on pp. 150, 155, and 158-159: Reprinted, by permission, from M. Angrosino, 1998, *Opportunity house: Ethnographic stories of mental retardation* (CA, USA: AltaMira Press), 40, 41, 97, 101, 103, 265, 266.

Brown text extracts on pp. 137 and 138-143: Reproduced from Brown, Leann "Boys' training": The inner sanctum. In Hickey, C., Fitzclarence, L. and Matthews, R. (eds.), *Where the boys are* with permission from University of New South Wales Press.

Denison text extracts on pp. 163-165 and 179-180: J. Denison, *Qualitative Inquiry* (2 (3)), pp. 352-353, 360, copyright (c) 1996 by Sage Publications, Inc. Reprinted by Permission of Sage Publications, Inc.

Denzin and Lincoln text extracts on pp. 4-5 and 6: N. Denzin and Y. Lincoln, *Handbook of qualitative research* (Second edition) pp. 12, 14-17, Copyright © 2000 by Sage Publications, Inc. Reprinted by Permission of Sage Publications, Inc.

Dowling Naess text extract on pp. 120-123: Reprinted, by permission, from F. Dowling Naess (1998), *Tales of Norwegian physical education teachers: A life history*. An unpublished PhD thesis, The Norwegian University of Sport and Physical Education, Oslo, Norway.

Duncan text extracts on pp. 85, 86, and 87: Reprinted, by permission, from M. Duncan, 2000, "Body as memory," *Sociology of Sport Journal* 17 (1): 60-64, 66-67.

Golden-Biddle text extracts on p. 47: K. Golden-Biddle and K. Locke, *Composing qualitative research*, pp. 3-4, 25-26, 29, 36-37, 60, Copyright © 1997 by Sage Publications, Inc. Reprinted by Permission of Sage Publications, Inc.

Jackson text extract on pp. 116-117: Reprinted, by permission, from D. Jackson, 1999, "Boxing glove" (poem). In *Talking bodies: Men's narratives of the body and sport*, edited by A. Sparkes and M. Silvennoinen (Jyvaskyla: SoPhi), 48.

Pelias text extracts on pp. 185 and 210-211: *Writing Performance: Poeticizing the Researchers Body* by Ronald J. Pelias © 1999 by the Board of Trustees, Southern Illinois University, reprinted by permission of the publisher.

Richardson text extracts on pp. 11, 15, 30, 129, 203, and 208-209: L. Richardson, *Handbook of qualitative research* (Second edition), pp. 923-925, 928-929, 931, 933-934, 936-937, 942, Copyright © 2000 by Sage Publications: London. Reprinted by Permission of Sage Publications, Inc.

Richardson text extracts on pp. 109-110 and 225: L. Richardson, *Handbook of interview research*, pp. 877, 881-883, 888, Copyright © 2001b by Sage Publications, Inc. Reprinted by Permission of Sage Publications, Inc.

Sparkes text extract from pp. 168-174: Reprinted, by permission, from A.C. Sparkes, 1997a, "Ethnographic fiction and representing the absent other," *Sport, Education and Society* 2 (1): 25-40. By permission of Taylor and Francis Ltd, **http://www.tandf.co.uk**

Swan text extract on pp. 118-119: Reprinted, by permission, from P. Swan, 1999, Three stages of aging. In *Talking bodies: Men's narratives of the body and sport*, edited by A. Sparkes and M. Silvennoinen (Jyvaskyla: SoPhi), 44-45.

Tsang text extracts on pp. 81, 82, 83, 84, 93-94, and 95: Reprinted, by permission, from T. Tsang, 2000, "Let me tell you a story," *Sociology of Sport Journal* 17 (10):44-59.

Van Maanen text extracts on pp. 41, 46-47, 50, 56, 60-61, and 212: Reprinted, by permission, from J. Van Maanen, 1988, *Tales of the field: On writing ethnography* (Chicago: University of Chicago Press), 46-47, 49-51, 53-54, 73-76, 79, 81, 92-93, 134-136. The University of Chicago Press, Chicago 60637. The University of Chicago Press, Ltd., London. © 1988 by The University of Chicago All rights reserved. Published 1988.

Wood text extract on pp. 175-177: Reprinted, by permission, from M. Wood, 2000, "Disappearing," *Sociology of Sport Journal* 17 (1):101-102.

Chapter 1

Surveying the Landscape

In 1987, I had my PhD viva. For the previous nine months I had been "writing up" my thesis, having completed a two-year ethnographic study of a group of physical educators involved in a teacher-initiated innovation. The end product was grandly titled *The Genesis of an Innovation: Emergent Concerns and Micropolitical Solutions.* Looking back, I now realize I wrote a *realist* tale. It worked—I passed. At the time, however, I was not aware that I was writing any kind of tale at all or that I had any choice of genres. At no point in the process was I asked to reflect on writing as a way of knowing or to reflect on the representational and ethical dilemmas inherent in the act of writing about other people and their lives from my own position as a situated author. No one else raised such questions, and I didn't either. I didn't need to. As I understood things back then, you did your ethnography, case study, interview project, or what have you; read around the area; analyzed the data; and then retired to your room to engage in the mysterious process of "writing up." To put it bluntly, writing in qualitative

research was not, at that time, an issue on the agenda. It was a dark secret, a mystery.

> Writing is the dark secret of social science. A blank page, so full of possibilities, so rarely discussed—at least until recently. While many manuals pile up their instructions on how to do research—to interview, to experiment, to sample, to data analyze—typically very little is said about how "findings" get communicated to an audience. It is assumed. . . . And yet it is also a mystery.
>
> *Plummer, 2001, pp. 168-169*

Things have changed. In recent years, there has been, depending on your perspective, a paradigm war, a revolution, or at the least a major upheaval in the social sciences. As part of this turbulence, there has been an outpouring of texts that have not only focused on how researchers have traditionally written about themselves and the social world they investigate but also suggested ways in which this might be done differently.[1] Indeed, in some cases, *very* differently. New journals, such as *Qualitative Inquiry*, launched in 1995, have emerged that actively encourage and pioneer new forms of writing in a variety of disciplines. Likewise, other well-established journals, such as *The Sociological Quarterly, International Journal of Qualitative Studies in Education, Journal of Contemporary Ethnography, Qualitative Sociology*, and *Qualitative Studies in Psychology*, have begun to accept qualitative articles written in different styles. Indeed, in the editorial to the first edition of *Qualitative Research*, published in April, 2001, P. Atkinson, Coffey, and Delamont state that "good papers couched in traditional forms and in radical styles will be published. . . . The sort of alternative representational modes we have in mind include: a dialogic approach; ethno-drama or ethno-theatre; and poetry" (pp. 12-13). Perhaps, not surprisingly, Punch comments:

> The rethinking of research that has accompanied both the paradigm debates and the emergence of new perspectives has included research writing: how the research is to be put into written forms, and communicated. . . . this rethinking has brought awareness of the choices about writing, identifying the conventional quantitative writing model as just one of the sets of choices. The appreciation of a

---

[1] For examples see Angrosino, 1998; P. Atkinson, 1990, 1991, 1992; Barone, 2000; Bochner, 1997, 2001; Bochner & Ellis, 2002; Denzin, 1997; Ellis, 1993, 1995a, 1995b, 1995c, 1998, 2001; Ellis & Bochner, 1996; Ellis & Flaherty, 1992; Golden-Biddle & Locke, 1997; Hertz, 1997; L. Richardson, 1990, 1994, 1997, 2000, 2001a, 2001b; Tierney & Lincoln, 1997; Van Maanen, 1988, 1995a, 1995b; Wolcott, 1990, 1994, 1995; Woods, 1999.

wider range of choices has meant a freeing up of some of the restrictions about writing, and is encouraging experimentation with newer forms of writing. As a result, there is a proliferation of forms of writing in qualitative research, the older models of reporting being mixed with other approaches.

Punch, 1998, p. 266

There is now talk of confessional tales, autoethnographies, poetic representations, ethnodrama, and ethnographic fiction as alternative and legitimate ways for qualitative researchers to represent their findings. But how has this situation come about, and what does it all mean? A whole book could be devoted to answering these two questions, so my aims in an introductory chapter can be modest. Accordingly, in this chapter I present a highly selective view of the shifting social science landscape of recent years to signal why the issue of representation has moved from the background to the foreground. I then briefly consider the implications of this foregrounding for qualitative researchers. This introduction provides a context for the other chapters in this book, which take you on a journey into different writing practices as ways of knowing about the world of sport and physical activity.

The Shifting Landscape: Momentary Glimpses

In the first edition of the *Handbook of Qualitative Research*, Denzin and Lincoln (1994) provided a useful way to review the history of qualitative research in North America during the 20th century. Their approach focused on five key *moments* in this history. In the second edition (2000), they developed their approach and spoke of seven moments. These can be summarized as follows.

First Moment

The *traditional* period, or moment, began in the early 1900s and continued until World War II. In this period, qualitative researchers aspired to "objective" colonizing accounts of field experiences shaped in the image of the positivist scientist paradigm: "They were concerned with offering valid, reliable, and objective interpretations in their writings. The 'other' who was studied was alien, foreign, and strange" (Denzin & Lincoln, 2000, p. 12).

Second Moment

The *modernist* phase extended through the postwar years to the 1970s and is still present in the work of many researchers. In this period, influential texts attempted to formalize qualitative methods with a view to making qualitative research as rigorous as its quantitative counterpart: "Thus did work in the modernist period clothe itself in the language and rhetoric of positivism and postpositivist discourse" (Denzin & Lincoln, 2000, p. 14). This was, however, a moment of creative ferment as qualitative researchers attempted to study important social processes, such as deviance and social control in the classroom and society.

Third Moment

The third stage (1970-1986) was the moment of *blurred genres*. By the beginning of this stage, qualitative researchers had a full complement of paradigms, methods, and strategies to employ in their research. Various formats for reporting research were in use, and diverse ways of collecting and analyzing empirical data were also available. Leading scholars proposed that the boundaries between the social sciences and the humanities had become blurred. The boundaries softened between different arts, between science and art, between fact and fiction, and between different academic disciplines so as to make possible the use of writing styles and genres that would previously have been considered inferior or nonliterary:

> At issue now is the author's presence in the interpretive text. How can the researcher speak with authority in an age when there are no longer any firm rules concerning the text, including the author's place in it, its standards of evaluation, and its subject matter?
>
> *Denzin & Lincoln, 2000, p. 15*

During this period, the naturalistic, postpositivistic, and constructivist paradigms gained power, especially in education, and several qualitative journals were in place by the end of the 1970s.

Fourth Moment

During the mid-1980s a profound rupture occurred, which signaled the arrival of a *crisis of representation*. This crisis focused attention on the writing practices of researchers. According to G. Marcus and Fischer (1986), the crisis arose from a growing uncertainty about the adequate means to describe social reality, which led to a reassessment of dominant ideas across the human sciences. Here, it was not just the ideas themselves that came under attack but also the paradigmatic style in which they were represented—how we "write" (explain, describe, index) the social:

> New models of truth, method, and representation were sought. . . . Critical, feminist, and epistemologies of color now competed for attention in this arena. Issues such as validity, reliability, and

objectivity, previously believed settled, were once more problematic. Pattern and interpretive theories, as opposed to causal, linear theories, were now more common, as writers continued to challenge older models of truth and meaning.

Denzin & Lincoln, 2000, p. 16

The crisis of representation remains with us today in the form of a continued questioning of the assumption that qualitative researchers can directly capture lived experience. Such experience is now taken to be *created* in the social text by the researcher, which means that the link between text and experience has become increasingly problematic. What is clear is that the crisis of representation moves qualitative research in new and critical directions. It also acts as a trigger point for the ongoing *crisis of legitimation,* which questions the traditional criteria used for evaluating and interpreting qualitative research. Involved here is a serious rethinking of terms such as validity, generalizability, and reliability: "The crisis asks, how are qualitative studies to be evaluated in the contemporary poststructural moment?" (Denzin & Lincoln, 2000, p. 17). Finally, in tandem, the dual crises of representation and legitimation feed into a third, which is the *crisis of praxis.* This crisis asks, "Is it possible to effect change in the world if society is only and always a text?" (p. 17).

Fifth Moment

The struggles to make sense of the triple crises described above shaped the postmodern and poststructuralist period of *experimental writing* that constitutes the fifth moment. Denzin and Lincoln signaled this period as beginning in the early 1990s. Here, new ways of composing ethnography were explored, theories were read as tales from the field, and writers struggled with different ways to represent the "other." New representational concerns emerged:

Epistemologies from previously silenced groups emerged to offer solutions to these problems. The concept of the aloof observer has been abandoned. More action, participatory and activist-orientated research is on the horizon. The search for grand narratives is being replaced by more local, small-scale theories fitted to specific problems and particular situations.

Denzin & Lincoln, 2000, p. 17

Sixth and Seventh Moments

According to Denzin and Lincoln, the sixth *(postexperimental)* and seventh *(future)* moments are on us:

Fictional ethnographies, ethnographic poetry, and multimedia texts are today taken for granted. Postexperimental writers seek to connect their writings to the needs of a free and democratic society. The demands of a moral and sacred qualitative social science are actively explored by a host of new writers from many different disciplines.

Denzin & Lincoln, 2000, p. 17

Indeed, Banks and Banks (1998) note that for at least the last 20 years anthropologists, sociologists, and others (such as literary critics, creative writers, and reader-response theorists) have been experimenting and theorizing about authorial voice, considering the nature and role of the "other," and reflecting on narrative form and authority.

The Problem With Moments

In reflecting on their "moments" approach, Denzin and Lincoln (2000) draw four conclusions, noting that it is, like all histories, somewhat arbitrary. First, they point out that each of the earlier historical moments they identified is still operating in the present, "either as a legacy or a set of practices that researchers continue to follow or argue against" (p. 18). Second, they argue that an embarrassment of choices now characterizes the field of qualitative research in that there have never been so many paradigms, strategies of inquiry, or methods of analysis to draw on. Third, they suggest that "we are in a moment of discovery and rediscovery, as new ways of looking, interpreting, arguing, and writing are debated and discussed" (p. 18). Finally, they assert that the qualitative research act can no longer be viewed from within a neutral or objectivist positivist perspective: "Class, race, gender, and ethnicity shape the process of inquiry, making research a multicultural process" (p 18).

Problems, however, remain with this moments approach. First, as Denzin and Lincoln (2000) acknowledge, they focus on North America, and the moments they describe might not apply to other countries. Second, they tend to focus on the discipline of sociology and anthropology. Other disciplines might not have proceeded at the same rate or have "progressed" so far in terms of moments. For example, Delamont, Coffey, and Atkinson (2000) illustrate the difficulties of applying this moments approach, and its time periods, to educational research.

The time span of the moments described by Denzin and Lincoln (2000) is difficult to apply to sport and physical activity where qualitative forms of inquiry remain relatively new kids on the block. For example, the third moment, the moment of blurred genres between 1970 and 1986, encompasses the period in the 1980s when debates about qualitative research first began to emerge and qualitative articles first began to get published in journals associated with sport and physical activity (Sparkes, 1992). Furthermore, since its arrival on the scene, qualitative research within the various subdisciplines of sport and physical activity has developed at different rates.

For example, in terms of the issue of validity, I have suggested that qualitative researchers in sport psychology during the late 1990s were operating in what Denzin and Lincoln (1994, 2000) describe as the second, "modernist" moment, and that the moments of blurred genres and the crisis of representation had yet to make their mark (Sparkes, 1998c). In contrast, in sport sociology and physical education, there are signs that during the early 1990s the fourth moment, which heralded the crisis of representation and legitimation along with the influence of postmodernism, began to touch these subdisciplines (see Bain, 1992; Cole, 1991; Evans, 1992; Foley, 1992; Gore, 1990; Lyons, 1992; Sparkes, 1991, 1995; Tinning, 1992) and has continued to touch them with increasing urgency in recent years (see Fernandez-Balboa, 1998; Kelly, Hickey, & Tinning, 2000; Markula, Grant, & Denison, 2001; Nilges, 2001; Oliver, 1998; Sparkes, 2000; Tsang, 2000; Wright, 1995, 1996, 2000).

Of particular significance here is the special issue of *The Sociology of Sport Journal* in 2000 that focused on imagining sociological narratives. In their introduction, Denison and Rinehart, the guest editors, pointed out that the motivations behind the special issue were to create a space for sport sociologists who have turned to more evocative ways of writing and to help legitimize the use of fiction and stories as nuanced ways to write up experimental ethnography. It is too early to assess the impact of this special issue on the sociology of sport research community and to see if this community warms to the notion of fiction as a legitimate form of inquiry. That it took a special issue of this journal to raise such an issue, however—that the editors knew its contents would be contentious and spark heated debate—indicates another problem for the periodization of the moments approach. Denzin and Lincoln (2000) claim that the sixth moment, the postexperimental, is already with us in the disciplines they speak of, where fictional ethnographies, ethnographic poetry, and multimedia texts are today taken for granted. This is not the case in sport and physical activity. Indeed, if it were, there would be little point in my writing this book!

Finally, concerns have been expressed about the succession of moments approach used by Denzin and Lincoln (1994, 2000) and Denzin (1997). For example, Anderson feels that as informative and insightful as their taxonomic analysis is, it is too clean. Anderson suggests we are dealing not so much with succession as with diversification. For him,

> The development of new ethnographic moments or genres does not seem to signal the demise of previously existing ones, but rather adds

more options in the styles and analysis available to qualitative researchers. New genres proliferate, vying with earlier ones, rather than displacing them.

Anderson, 1999, p. 453

Anderson believes that ethnography has never been a monolithic enterprise and there have always been differences and tensions both within and between traditions. Accordingly, he warns against romanticizing the past as a time of more clear-cut and consensual standards.

Coffey and Atkinson (1996), Coffey (1999), and Van Maanen (1995a) also make the point that the representation of ethnography through textual production has always been contested and varied. Thus, qualitative writing has always reflected a variety of sorts and encompassed different styles plus textual conventions. In view of this, P. Atkinson, Coffey, and Delamont (1999) suggest, it is easy for such mappings as the moments approach, which attempt to periodize the development of qualitative research, to gloss over the historical persistence of tensions and difference by too neatly packaging certain moments. As a consequence, these moments, as part of a developmental narrative, become ossified in ways that deflect critical scrutiny and evaluation. In contrast, P. Atkinson et al. suggest that rather than using the temporal metaphor of *moment* to describe the historical development of the ethnographic field, it might be more appropriate to speak of *vectors*, "implying the directionality of forces in an intellectual field" (p. 465).

Even those who advocate a vectors approach, however, acknowledge that the conventions of ethnographic representation and authorship have recently undergone further diversification, change, and contestation. For example, Coffey states:

> Despite a recognition that the production of the ethnographic text has never been wholly static or monolithic, it has only been of late that the production of texts and the reading of these texts have been the subject of detailed, critical and self-conscious scrutiny.

Coffey, 1999, p. 141

Coffey identifies various temporal and textual movements across a variety of disciplines as influencing this intense scrutiny. These include postcolonialism, postmodernism, feminism and postfeminism, lesbian and gay studies, queer theory, disability studies, and critical race theory. Coffey also notes that, despite its weaknesses, the narrative account of Denzin and Lincoln (1994) identifies ethnography and qualitative research as encapsulating several perspectives, debates, and ongoing tensions.

In short, whether we think in terms of moments or vectors, *something* has happened to qualitative research in recent years that has foregrounded issues of representation and legitimation and loosened the grip of specific styles of writing within the social science community.

The Crisis of Representation

This crisis asks the questions, Who is the "Other"? Can we ever hope to speak authentically of the experience of the Other, or an Other? And if not, how do we create a social science that includes the Other?

Lincoln & Denzin, 2000, p. 1050

According to Van Maanen, there once was a time, a dream time, when ethnography was read as a straightforward cultural description based on the firsthand experience an author had with a group of people that was strange to both the author and the reader:

One simply staked out the group, lived with them for a while, took notes on what they said and did, and went home to write it all up. If anything, ethnography looked like a rather pleasant, peaceful, and instructive form of travel writing.

Van Maanen, 1995a, p. 1

As P. Atkinson (1990, 1992) and Wolcott (1990) remind us, however, ethnography refers to both a research *process* and a textual *product*. Although the two are intimately related, the former ordinarily involves original fieldwork and requires the organization and editing of material for presentation, while the latter refers to the presentation itself, which ordinarily takes its form in prose. This relationship is evident in the three activity phases associated with ethnography, each of which, Van Maanen (1995a) suggests, raises distinct and problematic concerns for the subjects, the producers, and the consumers of the work. Phase 1 concerns the collection of information, or data, on a specified (or proposed) culture; that is, fieldwork. Phase 2 refers to the construction of an ethnographic report, or account, and, in particular, to the specific compositional practices the ethnographer uses to fashion a cultural portrait; that is, "writing it up." Finally, phase 3 occurs with the reading and reception of an ethnographic text across various audience segments.

In the past, Agar (1995) argues, phase 2 (the product) has suffered from benign neglect. Indeed, P. Atkinson talks of "a collective amnesia concerning literary modes of representation" (1991, p. 165). This was, and still is for many, partly induced by qualitative researchers' desire to gain academic respectability by distancing themselves from the more overtly literary aspects of their work. These aspects, however, have become increasingly difficult to ignore.

As Geertz noted, the gap between engaging others where they are and representing them where they aren't, always immense but not much noticed, has suddenly become extremely visible: "What was once technically difficult, getting 'their' lives into 'our' works, has turned morally, politically, even epistemologically, delicate" (1988, p. 130). This shift leads Van Maanen (1995a) to argue that the "dream time" is over. For him, the cultural representation business has become quite tricky, and just what is required of ethnographers now is by no means clear. Indeed, Van Maanen notes that among the producers and consumers alike, restlessness is the norm, and the master trope these days appears to be *J'Accuse!*

According to Agar, this has led to a situation now where "the ethnographic text will never again be taken for granted" (1995, pp. 112-113). Textual issues such as conventions of representation, along with genre constraints such as the distribution of narrative and descriptive passages, "have been converted from out-of-awareness tradition into matters of conscious debate" (p. 73). Similarly, Clifford (1986), in the introductory chapter to a book significantly titled *Writing Culture: The Poetics and Politics of Ethnography*, focuses specifically on the issue of writing—the making of texts. For him this activity is no longer a marginal, or occulted, dimension but is central to what anthropologists do both in the field and thereafter. "The fact that it has not until recently been portrayed or seriously discussed reflects the persistence of an ideology claiming transparency of representation and immediacy of experience. Writing reduced to method: keeping good field notes, making accurate maps, 'writing up' results" (p. 2). Not surprisingly, in recent years increased attention has been given to the process of writing as it transports the researcher from the "context of discovery" to the "context of presentation" (Plath, 1990, p. 376). The *product* of inquiry—the written account—is now under intense scrutiny.

We live in unsettled times because, as Tierney explains, what social scientists thought they knew, they no longer know: "What one assumed was the correct way to present data is no longer accepted without question; even the idea of *data* raises postmodernist eyebrows where one questions the meaning of what makes a fact a fact, what makes a text a text" (1999, p. 307). Similarly, Lather places both postmodernism and poststructuralism under a *deconstruction* paradigm defined as "to keep things in process, to disrupt, to keep the system in play, to set up procedures to continuously demystify the realities we create, to fight the tendencies for our categories to congeal" (1991, p. 13). This unsettling is welcomed by some. For example, L. Richardson believes we are fortunate to be working in a postmodern climate, as this is a time when a multitude of approaches to knowing and telling exist side by side.

> The core of postmodernism is the *doubt* that any method or theory, discourse or genre, tradition or novelty, has a universal and general claim as the "right" or the privileged form of authoritative knowledge. Postmodernism *suspects* all truth claims as masking and serving particular interests in local, cultural, and political struggles. But it does not automatically reject conventional methods of knowing and telling as false and archaic. Rather, it opens those standard methods to inquiry and introduces new methods, which are also, then subject to critique.
>
> *L. Richardson, 2000, p. 928*

The doubt inherent in postmodernism, therefore, distrusts all methods and forms of representations equally and seeks to continually destabilize knowledge hierarchies.[2] But, as Tierney and Lincoln (1997) and L. Richardson (2000) are quick to point out, the postmodernist position does allow us to know *something* that is partial and situated without us having to claim to know everything, or suggesting that everything that has gone before is "false" or that the views of scholars a generation ago were "wrong." As Nilges (2001) argues, in the postmodern, there are multiple positions from which to know and standards of truth are always partial, context dependent, and embedded in webs of power relationships. For her, in this era, qualitative researchers writing about social life must grapple with the epistemological challenge that knowledge can never be understood and textualized from a fully objective sphere. This is because the researcher is, in part, a product of the social context and processes being studied. Knowledge, therefore, is not only historically and contextually bound but is actually constructed through a process of reflexive mediation, where the world that is studied is created, in part, by the author's experience and the way the text is written.

It is now recognized that writing is an integral feature of the research enterprise whereby our findings are inscribed in the way we write about

---

[2] It is beyond the scope of this chapter to engage in a detailed discussion of the characteristics of postmodernism and poststructuralism. For more in-depth coverage see Cherryholmes (1988), Featherstone (1991), Harvey (1990), Hollinger (1994), D. Lyon (1994), and Sarup (1989). The literature suggests that we should talk in the plural rather than the singular when addressing these domains, that is, of postmodern*isms* and poststructural*isms*. It is also clear that these are contested domains that are open to continual critique. The point is that we do not have to buy totally into the postmodern or poststructural positions to learn something from them.

things. They are not detached from the presentation of observations, reflections, and interpretations. In short, it is now realized that there can be no such thing as a neutral, innocent report since the conventions of the text and the language forms used are actively involved in the construction of various realities.

Rhetoric, Fashionings, and Fictions

According to Nelson, Megill, and McCloskey,

> Scholarship uses argument, and argument uses rhetoric. The "rhetoric" is not mere ornament or manipulation or trickery. It is rhetoric in the ancient sense of persuasive discourse. In matters from mathematical proof to literary criticism, scholars write rhetorically. Only occasionally do they reflect on that fact.
>
> *Nelson et al., 1987, p. 3*

Across a variety of disciplines, it is now recognized that *how* research is presented is at least as important as *what* is presented. As part of this recognition, the classic issues of rhetoric, such as voice, style, and audience, have come to the fore. In an extensive exploration of how social scientists use literary and rhetorical conventions to convey their findings and arguments, P. Atkinson notes the following:

> It is difficult to think how *any* written or spoken text could convey "facts" or "findings," let alone analyses, hypotheses, conjectures, criticisms and refutations without recourse to conventionally appropriate textual formats. In the reader's evaluative readings, then, form and content are inextricably linked. The text cannot simply transcribe or report, but it must also persuade. . . . Indeed, sociological texts in general are inescapably rhetorical. Whether they adopt an explicitly exhortatory tone, or purport merely to report neutral "facts," they rely on devices of persuasion to construct plausible accounts, striking contrasts, historical inevitabilities; to link data into convincing sequences of cause and effect; to embed theory into data and vice versa.
>
> *P. Atkinson, 1990, pp. 15-16*

In keeping with the linguistic, or rhetorical, turn in social theory influenced by postmodernism and poststructuralism, experience is now taken to be created in the social text by the researcher. As L. Richardson points out, "Language does not 'reflect' social reality, but produces meaning, creates social reality. Different languages and different discourses within a given language divide up the world and give it meaning in ways that are not reducible to one another" (2000, pp. 928-929). Thus, the link between text and experience has become increasingly problematic. Language is not merely mimetic—words do not unproblematically contain meanings—and so language is no longer taken to be a transparent medium through which the world may be experienced and expressed. Similarly, neither speech nor writing are taken to furnish a privileged, neutral mechanism of representation. Rather, as P. Atkinson reminds us,

> Language is always "incomplete." It does not give us an exhaustive description of the physical or social world. However "factual" or "realistic" a text appears to be, it is inescapably dependent on the conventions of reading and writing that its producer and consumers bring to bear.
>
> *P. Atkinson, 1992, p. 38*

Language, therefore, is now seen to be a constitutive force that creates a particular view of reality. That is, it shapes us just as we shape it. Thus, the relations between language and "reality" are problematic and a site of exploration, contestation, and struggle. Rinehart (1998b) exemplifies this point when he reflects on the dilemmas he encountered when he tried to write about his experiences of the 26th annual Super Bowl in the United States:

> I was faced with a dilemma: the experience itself and my field notes and home video were like big blobs of paint on a blank canvas. They were unartful. They were non-narratives, with no apparent linearity, no dramaturgical sense. They occasionally ran together, once in a while made sense to me, but as a mishmash probably would be of no interest to scholars, the public, or even Alyssa and Nicholas, my two children. I couldn't see the stories I told as being cogent whatsoever. I had to chew on the experience a bit, reconstitute it into something that would make some kind of sense. The experience itself . . . required my mediation. . . . I would have to inscribe a pattern, a form, a written logic, upon the raw chronological but non-linear experience. By virtue of such tinkering, I would forever change the experience. . . . Narrational conventions shape the way writing is conveyed—and help to form how it is received.
>
> *Rinehart, 1998b, pp. 69-70*

Against this backdrop it is interesting to note the views of Geertz, who reminds us that all the research stories we produce are fictions, "fictions in the sense that they are 'something made,' 'something fashioned'—the original meaning of *fictio*—not that they are false, unfactual, or merely 'as if' thought experiments" (1974, p. 15). In relation to this, Strathern (1987) argues that any form of writing about research is a *persuasive fiction* that employs specific literary strategies. For Strathern, marking a piece of writing as "literary" is like marking out a person as having a "personality":

> Obviously, insofar as any piece of writing aims for a certain effect, it must be a literary production. . . . So whether a writer chooses (say) a "scientific" style or a "literary" one signals the kind of fiction it is; there cannot be a choice to eschew fiction altogether.
>
> *Strathern, 1987, p. 251*

In a similar vein, Bochner and Ellis note that, try as we may to report and represent accurately, "we necessarily invent and construct the cultures we write about. We cannot help but read something into what is there, because we are there with it" (1996b, p. 20). In support of this view, Coffey and Atkinson comment, "the boundaries between fact and fiction are sometimes difficult to draw at the best of times. All written work, however factual and authoritative, is composed and crafted" (1996, p. 127).

For some, any hint that the production of qualitative texts involves fictional techniques, fashioning, constructing, inventing, or composing is a source of anxiety and concern. Such anxieties, according to P. Atkinson (1992), arise only in relation to the most naive of belief systems:

> The inescapably textual character of "data" is an offence only if one clings to the view that there might be some other mode of presentation. But if it is recognized that there is no possibility of "literal" and unmediated apperception and recording, then many of the most threatening misgivings may be allayed.
>
> *P. Atkinson, 1992, p. 16*

For P. Atkinson, there is no great epistemological problem in the fact that neither field notes, nor transcripts, nor our final research reports are "literal" renderings. He argues that, recognizing our methods of reading and writing in ethnography to be thoroughly conventional and contrived, drawing on the same conventions as other literary forms (including fiction) is not a reason for undue anxiety or pessimism. Scholarship is not thereby vitiated, and the "recognition that all human inquiry and reportage are essentially the same is not a recipe for nihilism or a loss of scholarly standards" (p. 3).

Indeed, various scholars have suggested that there are many benefits to be gained from the more reflective stance towards texts initiated by the crisis of representation. For example, Coffey and Atkinson (1996) argue that in the current climate we should be more able to make disciplined and principled choices about how to represent and reconstruct social worlds and social actors, social scenes and social action. Making such choices directs how we write our accounts and directs us toward new ways of representing our research endeavors. With regard to making informed choices, Lincoln notes the following:

> The idea that we can think consciously about presenting and re-presenting the stories we tell proffers an enticing invitation to think reflexively and self-consciously—not just about the fieldwork we do, but also about the means we choose and use to relay our fieldwork tales to audiences. The choice implied in reflexivity leaves open the possibility that we can consciously take our narrative voice and reframe it. . . . The concept of choice, however, is a powerful one. Choice implies intention. Intention implies a kind of deliberation, and deliberation is at the center of our "story" here: we have choices, and those choices can and will reveal different intentions.
>
> *Lincoln, 1997, pp. 38-39*

As issues of writing style and genre become part of our methodological awareness, then other benefits emerge in relation to how we understand our data. As Coffey and Atkinson argue, "How we write is, effectively, an analytical issue. . . . It is important to think about the kind of written work desired" (1996, p. 117). They add, "We also wish to stress that decisions about representation are not optional extras. Any and all our modes of representation have significant implications for analysis" (p. 137). That is, modes of writing along with other forms of representation are fundamental to the work of qualitative data analysis. As a consequence, we can no longer relegate the production of our scholarly works to an apparently mechanical and minor aspect of the research, that is, the "writing up" stage. Quite simply, writing and representation cannot be divorced from analysis, and each should be thought of as analytic in its own right. In support of this view, L. Richardson argues as follows:

> Although we usually think about writing as a mode of "telling" about the social world, writing is not just a mopping-up activity at the end of the research project. Writing is also a way of "knowing"—a method of discovery and analysis. By writing in different ways, we discover new aspects of our topic and our relationship to it. Form and content are inseparable.
>
> *L. Richardson, 2000, p. 923*

Talking of his concerns and aspirations for qualitative research in the new millennium, Eisner points out that there has been a growing realization among researchers of something that artists have long known; namely, "that form matters, that content and form cannot be separated, that how one says something is part and parcel of what is said" (2001, p. 138). For Eisner, the form of representation that one uses has something to do with the form of understanding one secures. He notes that, once this idea penetrated the research community, the form used to inquire and to express what one had learned was no minor consideration: "This idea, the idea that different forms could convey different meanings, that form and content cannot be separated, has led to the exploration of new modes of research" (p. 139).

For Coffey (1999), the intense renewal of interest about representational issues in qualitative research has had several practical consequences:

- A more self-conscious approach to writing explores how meaning is composed from data and how the data is made meaningful to others.
- The articulation of the self in the products of field research has become a matter of critical reflection.
- The impersonal, all-but-invisible narrator status of the ethnographic author has been questioned.
- The dynamic nature of power relationships in field research and ethnographic production has been highlighted, reevaluated, and located alongside issues of authorship, authenticity, and responsibility.

It would certainly seem that representation in qualitative inquiry can no longer be taken-for-granted aspects of the research process and product. The dream time is over. This is particularly true in relation to voices in the text.

Reflexivity and Voice

In recent years, according to Gergen and Gergen, one of the emerging innovations in qualitative methodology has been the increasing importance that researchers have placed on reflexivity.

Here investigators seek ways of demonstrating to their audiences their historical and geographic situatedness, their personal investments in the research, various biases they bring to their work, their surprises and "undoings" in the process of the research endeavor, the ways in which their choices of literary tropes lend rhetorical force to the research report, and/or the ways in which they have avoided or suppressed certain points of view.

Gergen & Gergen, 2000, p. 1027

For Hertz (1997), the notion of reflexivity implies a shift in our understanding of data and its collection toward something that is accomplished through detachment, internal dialogue, and a constant scrutiny of *what I know* and *how I know it*. She suggests that reflexive ethnographers do not simply report "facts" or "truths" but also actively construct interpretations of their experiences in the field and then question how these interpretations came about: "The outcome of reflexive social science is reflexive knowledge: statements that provide insight into the workings of the social world *and* insight on how the knowledge came into existence" (p. viii).

Thus, researchers need to reflect on the political dimensions of fieldwork, the webs of power that circulate in the research process, and how these shape the manner in which knowledge is constructed. Likewise, they need to consider how issues of gender, nationality, race, ethnicity, social class, age, religion, sexual identity, disability, and able-bodiedness shape knowledge construction. These issues may affect interactions in the field; who gets studied and who gets ignored; which questions are asked and which are left unasked; how people are written in and out of accounts; and how "others" and the self of the researcher are represented. As L. Richardson points out, "Self-reflexivity brings to consciousness some of the complex political/ideological agendas hidden in our writing. Truth claims are less easily validated now; desires to speak 'for' others are suspect" (2000, p. 936).

Some suggest that researchers-as-authors need to indicate their positioning in relation to the research process and the other people involved (Coe, 1991; L. Richardson, 1992a, 1997). They also suggest that researchers engage in a self-reflexive analysis of the social categories to which they belong, since these enter into and shape what constitutes knowledge in any project. Consequently, for them, the author needs to be written *into*, and not out of, the text. Although many qualitative researchers seem happy to write narratives that situate the "subjects" of inquiry in culturally and historically specific locations, they appear less assured about recognizing that they as authors also write from specific historical and cultural locations. Such acknowledgments would openly challenge the commonly held view of the researcher-as-author as a disembodied and neutral voice, a universal human subject outside of history who is hermetically sealed off from social categories.

Indeed, Hertz (1997) argues that since the researcher is an active participant in the research process, it is essential to understand the researcher's location of self. In speaking of the *ethnographic self*, Coffey (1999) argues that fieldwork is a social setting, inhabited by embodied, emotional, physical selves, and that fieldwork helps to shape, challenge, reproduce, maintain, reconstruct, and represent our selves and the selves of others. Thus, fieldwork is personal, emotional, and identity-oriented: "In writing, remembering and representing our fieldwork experiences we are involved in processes of self presentation and identity construction" (p. 1). As part of this process, M. Fine (1994) asks researchers to think long and hard about how they work the hyphen in the self-other relationship in ways that both separate and merge personal identities with inventions of others, and about the ways in which they tend to speak "of" and "for" others while occluding themselves and their investments. Accordingly, M. Fine and Weis argue that

> We interrogate in our writings who *we* are as we coproduce the narratives we presume to collect. It is now acknowledged that we, as critical ethnographers, have a responsibility to talk about our own identities, why we interrogate what we do, what we choose not to report, on whom we train our scholarly gaze, who is protected and *not* protected as we do our work.
>
> *M. Fine and Weis, 1998, p. 25*

Issues of how researchers represent themselves and others in their texts become particularly acute when it comes to how, when, and whose voices are included in the final text. As Hertz notes, reflexivity encompasses voice,

> But voice focuses more on the process of representation and writing than upon the processes of problem formation and data gathering. . . . Voice is a struggle to figure out how to present the author's self while simultaneously writing the respondents' accounts and representing their selves. Voice has multiple dimensions: First, there is the voice of the author. Second, there is the presentation of the voices of one's respondents within the text. A third dimension appears when the self is the subject of the inquiry.
>
> *Hertz, 1997, pp. xii, xi-xii*

For Woods, "how to represent the 'voice of the other' is a considerable issue" (1999, p. 55). Likewise, M. Fine, Weis, Weseen, and Wong (2000) also emphasize that how to write the self of the author into the text is highly problematic. For example, they argue that simply inserting autobiographical or personal information about the author often serves to establish the authority of the researcher and can lead to texts in which the self has been sanitized. As M. Fine et al. point out, however, "flooding the text with ruminations on the researcher's subjectivities also has the potential to silence participants/'subjects'" (2000, p. 109).

Whose voices are included in the text; how they are given weight, interpreted, prioritized, and juxtaposed; and how events and scenes are described are not just textual strategies but *political* concerns that have *moral* consequences. That is, how we as researchers choose to write about others, and ourselves, has profound implications not just for how readable the text is but also for how the people it portrays are "read" and understood.

According to Clarke and Humberstone (1997), Hall (1996), Luke and Gore (1992), Olesen (2000), and Stanley and Wise (1993), the question of voice has been a continuing and worrisome problem for all feminist qualitative researchers. Indeed, the influence of critical race theory, queer theory, and disability politics has accentuated these concerns in recent years in terms of how participants' voices are to be heard, with what authority, and in what form (Gamson, 2000; Ladson-Billings, 2000; Wendell, 1996). On this issue, L. Richardson asks the following questions:

> How does one's writing reflect one's social privileges? What part of my biography, my process is relevant to text writing? How do I write myself into the text without being self-absorbed or unduly narcissistic? How can I write so that others' "voices" are not only heard but listened to? For whom should we write? What consequences does our work have for the people we study, and what are my ethical responsibilities for those consequences? These are not only my personal issues; they are ones that engage (enrage) both feminist and postmodernist researchers.
>
> *L. Richardson, 1992a, p. 108*

Dewar (1993) argues that, as part of a rethinking of feminist analyses of sport using standpoint theory, authors need to acknowledge the ways in which their interpretations and writing are embedded in systems characterized by structured inequalities and various forms of oppression. Her own analysis, which deconstructs the notion of the generic woman in sport, does just this. It problematizes the relationship between women as subjects, researchers, and authors, who can be differently privileged with regard to the social categories they occupy.

> Having privilege because of my white skin, Anglophone heritage, age, body size, class, physical ability, and Christian heritage influences how I do my work. The challenge I face is to understand how sexism and lesbian oppression are shaped by other forms of oppression. This means recognizing that there are times when my voice will be heard and listened to because of the privilege that I have, which in turn means considering if, why, and how I exercise that privilege. It also

means that there are times when I need to be silent, listen, and take responsibility for learning. As a first step this means that I must include and integrate different voices into my teaching and writing.

Dewar, 1993, p. 222

Issues of power, privilege, and the potential for exploitation in a research process characterized by systematic inequalities have been a central concern for scholars from a variety of disciplines (Ellsworth, 1992; Gore, 1992; Orner, 1992). This is particularly so for those who are committed to giving voice to people who are not usually heard because their points of view are defined as unimportant or difficult to access by those in power. For example, the voices of children; women; people with disabilities; members of minority groups; lesbian, gay, and bisexual persons; and lower participants in formal organizations often have their voices silenced because they are stigmatized by those operating in the mainstream culture. As part of their attempts to challenge the various forms of oppression that operate in our society, however, these same scholars have problematized the notion of "giving" voice to others in terms of the mechanisms of power and privilege implied in such an action. As Dewar comments,

> The issue is not who has a voice; we all have voices and speak with them in very different ways. The problem arises when we define our strategy against oppression as one that enables us to "give" certain groups of people a voice. What does it mean to give? What kinds of relationships does this imply? What kind of power and privilege is implied in the act of giving? What does this say about how voices are heard and interpreted?
>
> *Dewar, 1991, p. 75*

M. Fine (1994) and M. Fine and Weis (1998) share similar concerns about how, too often in the past, qualitative research projects have engaged in a process of othering in ways that fail to challenge the dynamics of oppression. Indeed, Orner (1992) suggests that "Student voice, as it has been conceptualized in work which claims to empower students, is an oppressive construct—one that I argue perpetuates relations of domination in the name of liberation" (p. 75). Riessman is also unhappy with the notion of giving voice to the experiences of another and prefers to think of research as a "chorus of voices, with an embedded contrapuntal duet. . . . Some voices will have to be restrained to hear voices from below, to create a particular harmony, but a different interpreter might well allow other voices to dominate" (1993, p. 16).

Such thoughts about how voices are staged draw on recent literary and feminist critiques that focus attention on the place of "selves" in texts about dominated others. They call on a language of criticism that speaks of voices, positions, silencing, erasure, and the ways in which narratives, texts, selves, and others are reproduced through the politics of domination. All this brings to the fore key questions regarding who speaks in the text; whose story is being told; and who maintains control over the narrative and, by implication, over the purposes to which the story is put. Accordingly, Lincoln (1993) notes that, to break the silence of the silenced, the products of research must go beyond traditional and conventional narrative forms. For her, there need to be multiple texts directed toward multiple readers so that "changes not only in form, but also [in] rhetoric, must accompany inquiries" (p. 40). Likewise, as Cole points out with regard to critical-feminist ethnographies that require symmetrical relationships between the researcher and the researched,

> It raises questions about the rhetorical conventions used to locate and frame subjects in the texts and how their stories can most usefully be represented. . . . In efforts not to exploit subjects, voices can be situated as central and positioned in ways that make them meaningful and productive. . . . Dispersed authority challenges illusions of single textual authority by representing other authoritative voices.
>
> *Cole, 1991, pp. 44-45*

Just how to achieve this dispersed authority and writing in of multiple voices is extremely problematic. For example, M. Fine (1994) provides examples of ways of writing that self-consciously interrupt and work *against* the process of othering. These include texts that insert "uppity" voices, stances, and critiques to interrupt master narratives. Fine suggests that these tend to be written by women of color, "situated at the intersection of race and gender oppression" (p. 75). In another kind of text, qualitative researchers dissect elites' construction of self and other; "this work enables us to eavesdrop on privileged consciousness as it seeks to peel Self off Other" (p. 75). Finally, Fine considers texts that press social research for social activism. Here, "researchers raise questions about the ethics of involvement and the ethics of detachment, the illusions of objectivity and the borders of subjectivity, and the possibilities of collaborative work and the dilemmas of collusion" (p. 75).

With regard to collaborative work, some have called for the production of jointly authored texts between researchers and participants as a way of promoting multivocal or polyphonic texts (Clifford, 1983; Lyons, 1992; Van Maanen, 1988). While supportive of this move, S. Smith (1993) has expressed some concerns over the effect of what she calls "rampant polyphony" in texts that attempt to capture for readers the differential experiences of an event and legitimate the stories of subjects from all cultural locations. If all the stories were given the same valence, credibility, and weight, the politics of the text would seem democratic and egalitarian. The incorporation of so many voices, however, would threaten to vitiate the impact of any specific voice, and the story of the event would threaten to consume the individual story. In this situation, Smith argues, the potential benefits of polyphonic interpretation give way before a commodified cacophony in which the reader ends up only with a kind of tourist experience. In support of this view, Hastrup also questions the "utopia" of plural authorship, which grants the subjects the status of authors:

> The purpose of ethnography is to speak *about* something for somebody; it implies contextualization and reframing. At the autobiographical level ethnographers and informants are equals; but at the level of anthropological discourse their relationship is hierarchical. It is *our* choice to encompass their stories in a narrative of a different order. *We* select the quotations and edit the statements. We must not blur this major responsibility of ours by rhetorics of "many voices" and "multiple authorship" in ethnographic writing.
>
> *Hastrup, 1992, p. 122*

More recently, Gergen and Gergen (2000), having noted the many variations that exist on the theme of multiple voicing, point to the following problem:

> One of the most difficult questions is how the author/researcher should treat his or her own voice. Should it simply be one among many, or should it have special privileges by virtue of professional training? There is also the question of identifying who the author and the participants truly are; once we realize the possibilities of multiple voicing, it also becomes evident that each individual participant is polyvocal. Which of these voices is speaking in the research and why? What is, at the same time, suppressed?
>
> *Gergen & Gergen, 2000, p. 1028*

Others have also argued that the researcher must take the unavoidable responsibility for the final text, which involves interpretation, evaluation, and judgment, and should not attempt to displace it onto the subjects themselves (Geertz, 1988; Gorelick, 1991; J. Marcus, 1992; Stacey, 1991). Indeed, for L. Richardson (1990), proposals to still the sociologist's voice as writer are in danger of rejecting the value of sociological insight and also imply that somehow facts can exist without interpretation. She suggests that there is no one right answer to the problem of speaking for others; that we must realize that writing, as an intentional behavior, is a site of moral responsibility; and that "We can choose to write so that the voice of those we write about is respected, strong and true" (p. 38).

Since there is no way for those who choose to write in this world to avoid deploying their power, L. Richardson (1990) argues for a merging of progressive and postmodernist thinking about authors and authority. This merging would counterbalance two impulses. The first, progressive impulse is to "give" voice to those who have been silenced and to speak for others even though the author might not necessarily belong to the constituencies in question. In contrast, the second, postmodernist impulse attempts to delete the author and to dismantle distinctions between fact and fiction in ways that can disempower those given a voice through progressive writing, by deconstructing their stories and undermining their grounds for authority. Such impulses need to be integrated, according to L. Richardson, by a progressive-postmodernist rewriting that recognizes that while all knowledge is partial, embodied, and historically and culturally situated, this fact does not mean that there is no knowledge, or that situated knowledge is bad.

> There is no view from "everywhere," except for God. There is only a view from "somewhere," an embodied, historically and culturally situated speaker. . . . Rather than decrying our sociohistorical limitations, then, we can use them specifically to ask relevant (useful, empowering, enlightening) questions. Consequently, the most pressing issue, as I see it, is a practical-ethical one: how should we use our skills and privileges? . . . As qualitative researchers, we can more easily write as situated, positioned authors, giving up, if we choose, our *authority* over the people we study, but not the responsibility of *authorship* over our texts.
>
> *L. Richardson, 1990, pp. 27-28*

Recently, Olesen (2000) has echoed these points and argued that all feminists, regardless of the tradition they work in, must attend to representation, voice, and text in ways that avoid replication of the researcher and instead display representation of the participants. As Olesen emphasizes, this issue is not resolved though the simple presentation of research findings in new and shocking ways. Rather, she insists, it speaks to the ethical and analytical dilemmas inherent in the intertwining of the researcher, the participant, and the mutual creation of data. Finally, Olesen makes it clear that "there can be no dodging the researcher's responsibility for the account, the text, and the voices" (p. 236). This view is supported by P. Atkinson (1992), who argues that although qualitative researchers must use the representational practices that are available to them, this is not a dispensation for irresponsibility: "On the contrary, just as the researcher must take responsibility for theoretical and methodological decisions, so textual or representational decisions must be made responsibly" (p. 52).

Reflections

The landscape I have described in this chapter is unstable, shifting, in flux, unpredictable, and perhaps, for some, dangerous. According to Denzin and Lincoln, qualitative research is defined "primarily by a series of essential tensions, contradictions, and hesitations" (2000, p. xi). For them, competing discourses circulate and move in several directions at the same time, and "this has the effect of simultaneously creating new spaces, new possibilities, and new formations for qualitative research methods while closing down others" (p. xiv).

L. Richardson also notes the possibilities for change in the shifting landscape of qualitative research: "It provides an opportunity for us to review, critique, and re-vision writing" (2000, p. 936). Having acknowledged the greater freedom we now have to present our texts in a variety of forms to different audiences, however, she recognizes how the dilemmas associated with the crises of representation and legitimation create different constraints and problems that need to be dealt with. Thus, the greater freedom qualitative researchers now have to experiment with textual form does not guarantee a better product: "The opportunity for writing worthy texts—books and articles that are 'good reads'—are multiple, exciting, and demanding. But the work is harder. The guarantees are fewer. There is a lot more for us to think about" (p. 936).

Conducting qualitative research has always been a difficult task. But L. Richardson is right: There is now a lot more for us to think about. The dream time that enveloped me when I did my PhD in the late 1980s is certainly over. Issues of representation, legitimation, reflexivity, and voice, to name but a few, now confront qualitative researchers in sport and physical activity throughout their projects. It is impossible to remain untouched by them, and there are no simple answers to any of the dilemmas posed by these issues.

In the face of the dilemmas described in this chapter and the inherent uncertainties of a postmodern age, it would be easy to become overanxious, sound the retreat, and then reinvent the dream time. Another, more positive approach, advocated by Tierney (1997), would be to acknowledge that we—as subjects, objects, participants, authors, and narrators—will often feel lost and ill at ease. But we should not ignore the discomfort or wish it away; rather, we should use it as a resource "to confront the issues of identity and representation and consider how we might develop texts that highlight the problematic worlds we study, our relationships to such worlds, and how we translate them (p. 34).

Gergen and Gergen (2000) also comment on the creative, innovative, and transformative potential of working within a matrix of uncertainty that continually encourages researchers to cross the boundaries of established enclaves. They suggest that if we can "avoid impulses towards elimination, the rage to order, and the desire for unity and singularity, we can anticipate the continued flourishing of qualitative inquiry, full of serendipitous incidents and generative expansions" (pp. 1042-1043).

Like Gergen and Gergen (2000), L. Richardson (2000), and Tierney (1997), I am drawn toward the creative potential generated by the tensions, contradictions, and hesitations that currently characterize qualitative inquiry in several disciplines. For me, this is a sign of vitality rather than deterioration. Consequently, my hope is that the chapters that follow in this book are able to contribute to the re-visioning and generative expansion of qualitative inquiry in sport and physical activity in the coming years.

Chapter 2

Scientific Tales

In a book concerned with writing in qualitative research, it might seem strange to have a chapter devoted specifically to traditional scientific writing. Even though I do not wish to devote much space to it, its inclusion is necessary for several reasons. First, the scientific tale is the dominant tale not only in the natural sciences but also in the social sciences. Therefore, it is the dominant tale in sport and physical activity.

> In the world of objectivist, positivist social science, writing generally parodies the style of the physical sciences; the tables, the findings, the tested hypotheses, simply speak for themselves and the exercise is simply one of *presenting* not *writing* "the findings." The style here is largely that of the external privileged reporter merely reporting what has been found.
>
> Plummer, 2001, p. 169

Second, precisely because this approach is so dominant, the sets of interests that inform it and the fundamental assumptions that it adheres to regarding the

nature of reality, the nature of knowledge, and how knowledge is communicated are taken for granted (Guba, 1990; J. Smith, 1989, 1993; Sparkes, 1989, 1992). Furthermore, the ways in which this kind of tale is constructed, and the ways in which these assumptions and interests work to persuade a specific kind of audience, are also too often not subjected to scrutiny. For example, each year I begin my undergraduate qualitative research course with a review of the scientific model of inquiry. It is recognizable to the students, and they feel comfortable with it as a way of knowing. On the other hand, when I ask them what philosophical assumptions (ontological and epistemological) and sets of interests underpin this model, I am greeted by blank looks.

When, later in the course, I focus on the scientific way of writing and ask what its conventions and rhetorical features are and how it works to persuade, I am again met by bewildered looks. Indeed, some students seem offended that I use the term *rhetoric* in relation to a scientific telling. *Rhetoric* is seen as a dirty word and invokes the qualifying adjectives of *mere* or *empty*. The students' view is that scientific writing, by definition, is not *rhetorical* and is *styleless*. This, of course, is not the case, as this short chapter will indicate. The scientific tale has a history and, like any other kind of tale, it adheres to certain conventions.

Learning the Form

Writing about quantitative research has typically been a straightforward matter, with conventional models and structures to guide the writer. According to Miles and Huberman (1994), the following sections make up a traditional scientific report:

- Statement of problem
- Conceptual framework
- Research questions
- Method
- Data analysis
- Conclusions
- Discussion

This format will be recognizable to many readers who might also be familiar with the conventional organization of a manuscript for reporting research as proposed by the *Publication Manual of the American Psychological Association* (4th edition, 1994): title page, abstract, introduction, methods, results, discussion, multiple experiments, references, appendix, and author note.

Chapter 2 of the *Publication Manual* is devoted to the expression of ideas and has sections on writing style, grammar, and guidelines to reduce bias in language. The introduction to this chapter makes it clear that the general principles and style requirements provided are designed to facilitate clear communication—the prime objective of scientific reporting. This is done by "presenting ideas in an orderly manner and by expressing yourself smoothly and precisely. By developing ideas clearly and logically and leading readers smoothly from thought to thought, you make the task of reading an agreeable one" (p. 23). Under a section on writing style it states, "The requirements are explicit, but alternatives to prescribed forms are permissible if they ensure *clearer communication*. In all cases, the use of rules should be balanced with good judgment" (pp. 23-24) (my italics). The *Publication Manual* states the following under the section on smoothness of expression:

> Scientific prose and creative writing serve different purposes. Devices that are often found in creative writing—for example, setting up ambiguity, inserting the unexpected, omitting the expected, and suddenly shifting the topic, tense, or person—can confuse or disturb readers of scientific prose. Therefore, you should avoid these devices and aim for clear and logical communication.
>
> *American Psychological Association, 1994, p. 25*

Even though these guidelines are recognizable to many, they are not "natural" ways of writing about the social world, nor are they a clear form of communication to all audiences. As Plummer (2001) points out, one consequence of scientific writing is to render the majority of such texts unreadable to all but the smallest of like-minded *cognoscenti*. Indeed, if this writing style was natural, then its conventions would not need to be institutionalized and encoded by professional bodies. That is, there would be little need for the 368 pages of the *Publication Manual*.

As Golden-Biddle and Locke (1997) emphasize, scientific writing is not a natural way of writing; it has to be learned. They note that few of us start out by writing the "academic speak" of our chosen profession or discipline: "Rather, the language and writing practices that symbolize the culture of science traditionally are transmitted across generations of academics" (p. 4). That is, we have to learn (and learn well) the disciplinary code of science in which "it is concluded that," and where only "the data" (not researchers) "suggest that." Furthermore, as Golden-Biddle and Locke point out, "if the writing of scientific style does not come naturally, then neither does the reading of it" (p. 4). Thus, this tale is constructed with particular purposes in mind, with a view to appealing to a specific kind of implied reader who, it is assumed, will understand and appreciate the conventions being called on.

Going Back in Time

The development and refinement of the scientific tale are embedded in a long, complex philosophical history that has seen heated debates concerning the relationship between language and knowledge—how people understand and convey their understandings to each other. As L. Richardson (2000) comments,

> Styles of writing are neither fixed nor neutral but reflect the historically shifting domination of particular schools or paradigms. Social scientific writing, like all other forms of writing, is a sociohistorical construction, and therefore, mutable.
>
> *L. Richardson, 2000, p. 925*

According to Zeller and Farmer (1999), the current debate on writing styles is grounded in the Attic-Asiatic debates of the first century B.C.E. in ancient Greece and Rome, with the Sophists, Plato, and Cicero as some of the central characters. Zeller and Farmer create links between this ancient history and the dreams held by the members of the British Royal Society (founded in 1660) of a plain language that could faithfully describe the natural world, and hence the phenomena that constituted the objects of scientific inquiry. This group aspired to, and believed it was possible to achieve, a language free of valuation, of expression and poetry, of context and rhetoric. At this time, the experimental method and inductive reasoning were identified as the proper methods to advance knowledge. This led Thomas Sprat, spokesman for the society, to warn experimenters in 1667 not to use expressive language:

> The ill effects of this superfluity of talking, have already overwhelm'd most other Arts and Professions. . . . Who can behold, without indignation, how many mists and uncertainties, these specious Tropes and Figures have brought on Knowledge? . . . of all the Studies of men, nothing may be sooner obtain'd, than this vicious abundance of Phrase, this trick of Metaphors, this volubility of Tongue, which makes so great a noise in the world.
>
> *As cited in Zeller & Farmer, 1999, p. 3*

In a brief review of the historical context of writing conventions, L. Richardson (2000) notes that from the 17th century onward, the world of writing was divided into two separate kinds: literary and scientific. The former was associated with fiction, rhetoric, and subjectivity, whereas the latter was associated with fact, plain language, and objectivity. Thus, "Fiction was 'false' because it invented reality, unlike science, which was 'true,' because it purportedly 'reported' objective reality in an unambiguous voice" (p. 925).

During the 18th century, assaults on literature intensified. During this period, L. Richardson (2000) notes that David Hume depicted poets as professional liars, while Jeremy Bentham proposed that the ideal language would be one without words, only unambiguous symbols. In such a climate it is not surprising that by the 19th century, literature and science stood as two separate domains. As L. Richardson argues, literature became aligned with "art" and "culture," was value laden, and used metaphoric and ambiguous language. In contrast, "given to science, was the belief that its words were objective, precise, unambiguous, noncontextual and nonmetaphoric" (p. 925).

For Zeller and Farmer (1999), the British Royal Society's wish to confine words to their strictly referential function remains with us today. They argue that these wishes can be found in the admonitions of those writing about research in the social sciences, and, more generally, in the prescriptions of the *Publications Manual* (American Psychological Association, 1994) as described earlier in this chapter. Indeed, as Golden-Biddle and Locke (1997) have pointed out, until the late 1970s most academicians considered scientific texts to be nonpersuasive. The general belief was that science was a special discourse operating outside the domain of rhetoric. As opposed to artists and novelists, who were thought to persuade through language, scientists were thought to persuade through logic and evidence (Gusfield, 1981).

According to L. Richardson (2000), however, as the 20th century unfolded, the relationships between social scientific writing and literary writing grew in complexity and "the presumed solid demarcation between 'fact' and 'fiction' and between 'true' and 'imagined' were blurred" (p. 926). Indeed, in recent years, as scholars have focused on the textual arrangements and the metaphored nature of scientific writing, the nonrhetorical view of scientific writing has been exposed as nonsense.

On Scientific Style

According to Bruner (1986), scientific explanations aim to be formal, logical, and, where possible, mathematical. Such explanations, for Bruner, are framed by what he defines as the paradigmatic or logico-scientific mode of thought. The prototypical example of paradigmatic writing, the experimental report, contains a formal sequence of reasoning involving theory, reasoning about theory, deduction of hypotheses, arguments about correspondence rules and observation statements, specific empirical predictions following from particular operations, and so on.

For many, this hypothetico-deductive structure would appear to be purely paradigmatic. However, as Plummer (2001) comments, this approach is "stuffed full of artifice, convention and literary devices that seek to persuade the reader" (p. 169). Indeed, Cheyne and Tarulli argue that this form of writing can be read as a narrative or story that both promotes and adheres to a particular world view:

> The report begins with the traditional prologue, called by its paradigmatic name, the ABSTRACT, which provides a summary of the action to be recounted. Next, the INTRODUCTION begins with a review or *exposition* of the current state of affairs and, along the way, reveals some *trouble*, the prototypical occasion for narration. Trouble, in this context, typically arises from a lack, breach, or crisis in our understanding of some aspect of the field. The "call" for redress requires a preparation for action. To meet this challenge a hypothesis is introduced that promises to resolve the lack, breach, or crisis. The hypothesis *risks* a prediction and hence sets up a dramatic tension that can be resolved in the narrative only by the dramatic activity of the *ordeal* of the experimental test in the METHOD section. The research report reaches a *climax* in the final statistical tests of significance reported in the RESULTS. The climax is followed, with the resolution of the crisis, by the *denouement* of the DISCUSSION in which the implications of the tests for the theory and for future empirical adventures is considered.
>
> *Cheyne & Tarulli, 1998, p. 4*

Cheyne and Tarulli (1998) suggest that, from a structuralist perspective, the experimental report may be understood as embodying an idealized narrative *fabula* (i.e., structural frame or theme) that may nonetheless accommodate an indeterminate number of *sjuzhets* (i.e., sequences of events, or plots). For Cheyne and Tarulli, both the form and the content of these stories "may be important for the probative value and the persuasive impact of the report" (p. 4). They further note how the evident paradigmatic style of the report adheres to the canons of logic and evidence and how in its "implicit narrative structure it tells a potentially compelling drama of a hypothesis exposed to the ordeal of possible disconfirmation in the ongoing adventure of science" (p. 4).

Besides a growing recognition that scientific writing is permeated by narrative features, the apparent absence of style in scientific tales is now acknowledged as a rhetorical device in its own right. That is, the "style of no-style" is the style of science. In reviewing the rhetoric of science, P. Atkinson (1990) notes that the "standard" style and format "does not literally *report* the process of discovery, but imposes a reconstructed logic. The emphasis is on an inductivist discovery of the facts, while the role of personal interests, or circumstances, is elided. Moreover, the scientific paper does not simply report; it *persuades*" (p. 43). That is, the use of passive voice and the third person, the strict separation of method from findings and findings from interpretation, the strict accounting of steps followed for data collection and analysis, and the recurring appeals to significance and validity are intended, according to Sandelowski (1994), to persuade the readers that what they are reading is science and *not* art, and that the findings reported are objective and uncontaminated by the heart and mind of the researcher.

> Moreover, what distinguishes art from science is not the presence of literary and rhetorical (persuasive) devices in the one and their absence in the other. . . . the novel and research report are both modes of representing reality, not of presenting reality itself. Scientists no less than artists attempt to persuade their audiences of the value/validity of their findings by employing speaking and writing strategies to stake their claims. . . . The conventional research report is no less a highly stylized art form than the novel or poem.
>
> *Sandelowski, 1994, p. 53*

Indeed, Plummer emphasizes that if researchers do not adopt the conventions of this style, then readers do not take them seriously as scientists: "For that is what science has to look like to persuade" (2001, p. 169). As Eisner comments:

> We talk about our *findings*, implying somehow that we discover the world rather than construe it. We say in our discussion, "it turns out . . ." implying that how things are is nothing for which we have any responsibility. We talk and write in a voice void of any hint that there is a personal self behind the words that we utter; [we refer to] "the author," "the subject," "the researcher," or, miraculously we somehow multiply our individuality and write about what "we" found. All these linguistic conventions are, paradoxically, rhetorical devices designed to persuade the reader that we, as individuals, have no signature to assign to our work.
>
> *Eisner, 1988, p. 18*

P. Atkinson (1990) argues that since no single, neutral, objectively given medium directly corresponds to the "facts of the case," then there are many possible ways in which a scientific paper might be constructed. To illustrate this point, Atkinson draws on the work of Law and Williams (1982), who focus on how a group of researchers works collaboratively to "package" a paper to make it as attractive as possible to a specific audience. The researchers try to array people, events, findings, and facts in such a way that the readers interpret them as true, useful, good work, and so on. That is, the researchers structure and juxtapose elements in their paper in such a way that interested readers find themselves compelled to interpret the findings in the manner desired by the authors.

Of course, as Law and Williams (1982) point out, no text can actually "compel," or determine reception, on the part of the reader, although the authors' intentions may be orientated to such an outcome. The main point, however, is that the text is in a process of creation and is packaged in such a way as to perform three main functions: (1) to tie together objects and fact; (2) to tie together people; and (3) to tie together objects, facts, and people in ways that are stylistically and grammatically acceptable. The end result of all this, according to P. Atkinson (1990), is that "scientific" papers construct accounts that are plausible and persuasive. Of course, they are plausible and persuasive to particular communities who hold to a particular set of views about the world, what constitutes "proper" research, and how it should be reported in particular contexts.

Given an infinite variety of ways in which the scientific report could be shaped, it is interesting to note that the dominant version is informed by a widely shared image of the person in science. According to Gergen (1999), this metaphor is that of the "mind as mirror." Here, the mind inside the head is subjective and the world outside is objective. Thus, it is believed that one is objective when private experience is the perfect reflection of the natural world.

Unfortunately, as Gergen (1999) and J. Smith (1989) make clear, there is no way to separate subject from object, no means of knowing "what is inside the mind" or determining whose mind is reflecting reality more or less accurately. Therefore, Gergen argues, objectivity cannot refer to a relationship between mind and world; rather, objectivity and scientific reality are achieved by speaking and writing in particular ways that adhere to the mind-as-mirror metaphor. Thus, according to Plummer, scientific writing "is a means to an end. Any discussions about 'how to do it' can hence be relegated to mere technical problems of technique—syntax, grammar, and . . . 'house style'" (2001, p. 169). Accordingly, Golden-Biddle and Locke note that "The writing of these reports generally is taken to be minimally expressive so that the discovered phenomena can be reflected as clearly as possible in the text. This is a 'windowpane' model of language" (1997, p. 3).

The windowpane model of language operates via a number of rhetorical tricks of the trade. Here are but a few that have been identified by P. Atkinson (1990), Gergen (1999), and L. Richardson (1991). For the intended audience of peers to be convinced, the scientist's contribution needs to be seen as essentially coincidental with the unfolding realization about the objective state of the world. The impression is that any other scientist in the same situation would have been led to the same conclusion. Essentially, in keeping with positivistic assumptions regarding epistemology and ontology (see J. Smith, 1989), this form of language has to create the impression of knower and known as separate entities. To this end, much use is made of powerful metaphors. L. Richardson has identified three in particular that are regularly used to remove the "datum" from the temporal and human practices that produce it so that human beings, both researchers and the others involved, are *metaphored* out of this world.

> First, is the grammatical split between subject and object, a wholly unnoticed metaphor for the separation between "real" subjects and objects. The metaphor is particularly powerful because it is part of our language structure. Second, empiricism views language as a *tool*. The empiricist world is fixed and available for viewing through the instrumentality of language, downplaying that *what* we speak about is partly a function of *how* we speak. Third, empiricism uses a *management* metaphor. Data are "managed," variables are "manipulated," research is "designed," time is "flow-charted," "tables" are "produced," and "models" (like toothpaste and cars) are "tested." The three metaphors work together to reify a radical separation between subject and object and to create a static world, fixed in time and space.
>
> *L. Richardson, 1991, pp. 8-9*

Such metaphors, operating in combination with the stripped-down, abstracted, detached form of language, the impersonal voice, and the statement of conclusions as propositions or formulae, involve a realist, or externalizing, technique that objectifies through depersonalization. According to Gergen (1999), it is important that the scientific writer employ *distancing devices*. These work to "call attention away from the agent and place the object(s) at a seeming distance" (p. 74). This is in contrast to *personalizing devices*, which call attention to the object as a private possession of the mind. Therefore, the scientist is likely to speak of "the apparatus" rather than "my sense of an apparatus." Such distancing techniques give the impression that the text is an inert, neutral representation of the world, and that it exists quite independently of the interests and efforts of the researcher, who is presented as a neutral and disengaged analyst.

For those who believe the mind functions like a mirror, according to Gergen (1999), to establish themselves as authorities on "what exists," authors must show that their mirrors were indeed in a position to reflect.

Therefore, the authors might signal their experiential presence early on in a scientific report by using personal pronouns (such as "I" or "we") or possessives ("my" or "our"). Because the experience of a single person is suspect, however, Gergen notes that it is often useful to show that "other mirrors reflect the same thing" (p. 75). Therefore, regardless of variations in procedures, samples, laboratory environment, or the time of the research, the scientific report is constructed to give the impression that the same reflection was found in all other mirrors. Thus, "Smith demonstrated that . . . , Brown corroborated it . . . , and Jones found the same effect" (p. 75). Gergen notes more subtle rhetorical devices that suggest the author's mirror has a special advantage and is possibly superior to others, perhaps even granting a God's-eye view of reality. This effect is most often accomplished by the use of impersonal pronouns: for example, "it was found."

Finally, Gergen (1999) illustrates the rhetorical features of scientific reports that seemingly signal the death of interfering passions. After all, he asserts, "The mirror of the mind achieves objectivity when there is no interference, when it possesses no defect that might 'distort,' or 'bias' the image produced by the world" (p. 75). To convince the reader that there are no "mirror effects," the author uses phrases that grant the world an active power to create the image instead of showing the image as a characteristic of the mirror itself. Phrases like "the findings suggest" contribute to this sense of the "real" world telling us its secrets. At the same time, the text must demonstrate the absence of internal states (emotions, motives, values, and desires) because passions distort reflections. The intrusion of emotions would subvert any notion of objectivity in a research community where, as Woolgar argues, "The scientist needs to be the trusted teller of the tale but, at the same time, should not be seen as intruding upon the object" (1988, p. 75). Such effects are produced by a variety of rhetorical strategies and conventions that inform the scientific tale.

Reflections

These rhetorical techniques produce tales that are very persuasive in Western cultures, and their value should not be underestimated. As indicated, however, they tend to provide what Geertz (1988) has called "author-evacuated texts," where the author is everywhere but nowhere, a disembodied and abstract presence. The voices of others tend to be present only in quantified form, as when oral responses to questionnaires are translated into researcher categories for statistical analysis and presentation.

Of course, to recognize and reflect on the rhetorical features of scientific tales is not to argue for abandoning that rhetoric. As Gergen comments,

> Rather, the rhetoric of reality plays a very important role in communities. Often it is vital in achieving trust, community, and results. For example, when space scientists use this rhetoric they ask their colleagues to trust that they are using the language in the same way and for the same purposes as the remainder of the community. They are "calling a spade a spade" in terms of the community's standards, and as a result humans can walk on the moon.
>
> *Gergen, 1999, p. 76*

The rhetoric of science and the scientific tale do have an important part to play in how we construct *one* way of acting in our world. This does not mean, however, that in sport and physical activity we should limit ourselves to what scientific forms of representation can describe and explain. As Eisner states,

> I have no issue to take with science, social or otherwise, but scientific frameworks do not exhaust the ways in which we experience the world or render our experience of it public. Bias, ironically, comes not only from commission, but also from omission. Science, like the arts, omits as well as includes. In that sense, all forms of representation are biased.
>
> *Eisner, 2001, p. 140*

Thus, as Gergen (1999) points out, problems arise when the scientific way of representing reality is taken to be the *only* way, and other legitimate forms of representation and ways of knowing are denied. Here, the merits of science are overshadowed by the specter of *scientism*. According to Habermas, "Scientism means science's belief in itself: that is, the conviction that we can no longer understand science as *one* form of possible knowledge, but rather must identify knowledge with science" (1971, p. 4). As such, the chapters that follow need to be seen as a rejection of scientism and not of science and scientific tales per se. As stated in the introduction to this book, the intention is not to replace dominant forms of telling totally, but to *displace* them so that other tales can make their own valuable and legitimate contributions to how we understand the world around us.

Chapter 3

Realist Tales

The philosophical assumptions and interests that drive qualitative forms of inquiry are different from those that inform research conducted in positivist and postpositivist paradigms (Guba, 1990; Lincoln & Guba, 2000; J. Smith, 1989, 1993; Sparkes, 1992, 1998c). It is therefore not surprising that qualitative researchers report their work differently, draw on different discourses, and use different rhetorical strategies to persuade readers that their accounts are authoritative. What is more, there is little uniformity in the way that qualitative researchers report their work. There are still few agreed-upon written formats, writing guidelines, or stylistic suggestions for neophyte qualitative researchers to emulate. Miles and Huberman confirm this view:

> The reporting of qualitative data may be one of the most fertile fields going; there are no fixed formats, and the ways data are being analyzed and interpreted are getting more and more various. As qualitative analysts, we have few shared canons of how

our studies should be reported. Should we have normative agreement on this? Probably not now—or, some would say, ever.

Miles & Huberman, 1994, p. 299

Given qualitative researchers' basic assumptions and interests, and the lack of an all-embracing style, the creation of authoritative texts is extremely problematic—particularly in relation to the issue of "voice(s)." For example, the discourse of the natural sciences tends to deny its "objects" a voice. As Woolgar comments:

> Although electrons, particles and so on are credited with various attributes, they are constituted as incapable of giving opinions, developing their own theories and, in particular for our purposes, producing their own representations. The natural science discourse thus constitutes its objects as quintessentially docile and can act upon them at will.

Woolgar, 1988, p. 80

In contrast, Woolgar (1988) notes, various traditions in the social sciences wish to grant the objects a voice (and refer to them as "subjects" or "participants"). This generates difficulties for the rhetorical constitution of distance: "In particular, in the discourse associated with interpretive social science, subjects/objects are granted the ability to talk back, have their own opinions and even to constitute their own representations" (p. 80).

One particular kind of tale has come to dominate qualitative research in sport and physical activity: the *realist* tale, as defined by Van Maanen (1988). To illustrate the conventions that shape this kind of tale, this chapter focuses on three pieces of qualitative research that deal with the personal experiences of injury in sport using inductive forms of analysis. All three have been published in reputable refereed journals that use the American Psychological Association format. The first article is my own (Sparkes, 1998a) and is titled "Athletic Identity: An Achilles' Heel to the Survival of Self." It involves a narrative analysis of one young, elite athlete, whose career was terminated by illness. The second is a psychological study by Gould, Udry, Bridges, and Beck (1997) titled "Coping With Season-Ending Injuries." Here, the authors seek to identify the coping strategies and factors thought to facilitate recovery in elite skiers in the United States who suffered season-ending injuries. The third, by Young, White, and McTeer (1994), is a sociological study titled "Body Talk: Male Athletes Reflect on Sport, Injury, and Pain." Its focus is on current and former Canadian male athletes with experiences of injury and on how, in ways that both challenge and reinforce dominant notions of masculinity, they learn to disregard risk of physical harm and to normalize pain and disablement as part of the sport experience.

By using these three articles as points of comparison, I hope to illustrate how the realist tale operates via similar conventions in different subdisciplines and how it works from case studies of one person to studies using larger samples.

Experiential Author(ity)

According to Van Maanen (1988), the most striking characteristic of realist tales is the almost complete absence of the author from most segments of the finished text. Only the words, actions, and (presumably) thoughts of members of the studied culture are visible in the text.

> The fieldworker, having finished the job of collecting data, simply vanishes behind a steady descriptive narrative. . . . Ironically, by taking the "I" (the observer) out of the ethnographic report, the narrator's authority is apparently enhanced, the audience worries over personal subjectivity become moot. . . . the narrator of realist tales poses as an impersonal conduit who . . . passes on more-or-less objective data in a measured intellectual style that is uncontaminated by personal bias, political goals, or moral judgments. A studied neutrality characterizes the realist tale. . . . The presence of the author is relegated to very limited accounts of the conditions of the fieldwork (its location, length, research strategies, entrance procedures, etc.).
>
> *Van Maanen, 1988, pp. 46-47*

My own biographical study of a young elite athlete called Rachael consists of 20 pages—18 pages of research findings and 2 pages of references. The "I" or the "me" are only present in 1 1/2 pages (pp. 648-649) as part of the methodology section. Here, the announcement of my own presence is necessary to inform the reader of how I came to meet Rachael as one of her university lecturers and personal tutor, how "informed consent" was negotiated for her involvement in the study, how the interviews were conducted, their number (9) and duration (18 hours in total), and how pieces of reflective writing by Rachael were used in the overall process. In these paragraphs, the "I" and the "me" often slip into a "we," and each operates to confirm that various ethical issues associated with this kind of project had been dealt with appropriately.

Only a few lines hint at my own autobiographical positioning and how this might have shaped the development of my engagement with Rachael and my analysis of her life.

As an active listener, therefore, I shared my experiences when I deemed it to be appropriate and when I was invited to do so. For example, I have my own personal experiences of once having inhabited a highly skilled, performing body that developed a chronic back pain problem that prematurely ended a promising sports career. Often, these experiences were shared with Rachael during our interviews as I reacted to issues raised by her. This sharing of stories and insights into the lives of each other reduced the distance between us as lecturer and student. . . . each of us became more vulnerable to the other due to the insider knowledge that was gained, and this helped us to develop a trusting relationship that formed the basis of our collaboration.

Sparkes, 1998a, pp. 648-649

My brief personal positioning is embedded in a section that deals with the interviewing process and the nature of the interaction between Rachael and myself. In such situations, issues of openness, honesty, and trust become paramount methodological concerns. Thus, the inclusion of some personal details about the author and how they were shared with the participant confirm both how, and that, the trusting relationship required in this form of inquiry was developed. The "I," the "me," and sometimes the "we" are also mentioned in a few paragraphs of the next section of the article, which deals with data analysis and interpretation. After this, the "I" disappears.

The article by Gould et al. (1997) consists of 19 pages of text and just over 2 pages of references, notes, and acknowledgments. The method section of the article takes up approximately 2 pages and is divided into sections on design, sample, interview and interview guide, and data analysis. Here, among other things, we learn that retrospective in-depth interviews were conducted with 21 elite North American skiers who had suffered season-ending injuries. An interview guide was employed and the skiers were encouraged to "feel free to voice both their positive and negative experiences. They were also informed that there were no right or wrong answers" (p. 383). These interviews were transcribed and then inductively analyzed using hierarchical content data analysis procedures to identify major themes and patterns of like categories so that a hierarchy of responses could be established that moved from specific to general levels.

Under the section on interview and interview guide, we are told that each skier participated in an in-depth interview lasting between 60 and 90 minutes. After this, there is a momentary reference to one of the authors: "These interviews were tape-recorded and conducted by the same individual (a 33-year-old female) who was trained in qualitative research methodology and who had experienced major knee injuries herself. An interview guide was used to standardize all interviews and minimize bias" (Gould et al., 1997, p. 382).

Nothing more is said about this author or any of the others in the rest of the article. So why this brief announcement of authorial presence? How does it work? At one level, it hints at some of the recognized benefits that accrue from having a member of the culture, an "insider," interview other members of the culture—for example, ease of access, plus the potential for more trusting relationships and the participants' willingness to "open up" to one of their own. At the same time, Gould et al. (1997) also seem to consider this insider perspective—the subjective experiences of the interviewer—a threat to the enterprise. As a consequence, the revelation of an authorial presence with some experience related to the research problem is counteracted by the use of an interview guide to tame researcher subjectivity.

The article by Young et al. (1994) consists of 17 1/2 pages of text with 3 1/2 pages of references and acknowledgments. One page is devoted to research approach. Here, we learn that 16 current and former male athletes from Canada were involved in semistructured interviews of one to two hours' duration, framed by some theoretical questions and propositions that emerged from a literature review conducted by the authors. During the interviews, the authors remained open to new conceptual ideas and to unanticipated construction of meanings by subjects. Young et al. emphasize that "Subjects were allowed to speak in their own terms and categories wherever possible, and systematic analysis of the transcribed interviews yielded a number of themes" (p. 180).

The authors also note that despite discovering surprisingly little variability in the way each of them "read" the transcribed interviews, they recognized that as with other qualitative work, their findings represented their own interpretations. Thus, because of the nature of the methodology, they could not assert with "complete certainty" that their interpretations were valid. They comment, however, that

> As three male, former university-level participants in the kinds of aggressive sports our subjects play (football and rugby, specifically), each of us having experience in the very limiting, often painful downside of sport . . . we feel confident that our decoding of the data is accurate.
>
> Young et al., 1994, p. 179

Having acknowledged the possibility of bias due to the researchers' having experiences similar to those of the participants, Young et al. (1994)—in contrast to Gould et al. (1997)—use the revelation of authorial presence to placate these worries for the reader. That is, Young et al. imply that, precisely because of their biographical positioning, they are better able to decode the data in an accurate fashion that closely matches the lived realities of the participants.

To summarize, in each of these three articles, the narrator(s) as a first-person presence is markedly absent. Where an authorial presence is announced, it comes in the methods section of the article and not in the other sections making up the main body of the text. As Cole (1991) comments, the author's voice in realist tales is usually set apart from the main text in prefaces, method sections, and footnotes to indicate that a dispassionate observer was there, saw, and knows, "asserting a contradictory 'disembodied' objective presence and 'experiential' authority. The researcher constructs and positions him/herself as a conduit through which an 'other' culture is seemingly symmetrically decoded and recoded" (p. 39). Furthermore, when announced, the authorial presence is *confirmatory* for the purposes of persuasion. That is, it confirms either by dissolving audience worries over personal subjectivity or by making the case that this subjectivity provides some methodological advantage.

Therefore, in realist tales, moves to enhance experiential author(ity) produce texts that tend to be dominated by "scientific" narrators who are manifest only as dispassionate, camera-like observers/listeners (Marcus & Cushman, 1982). As Young points out, this bodily withdrawal from the realms of events, or disembodiment in it, leaves behind the spoor of the voyeur: "Though the body vanishes, its perceptual apparatus, especially the eyes, remains. Ethnographic writing instantiates an invisible perceiver in a visible world" (1991, pp. 221-223).

In this sense, realist tales are strikingly similar to scientific tales because they construct authority and objectivity through the use of a passive voice so as to obscure and apparently distance the disembodied author from the data.

The Participant's Point of View

According to Van Maanen (1988), realist tales are characterized by extensive, closely edited quotations. These are used to convey to the reader that the views expressed are not those of the researcher but are rather the authentic and representative remarks transcribed straight from the mouths of the participants. He points out, "What, precisely, might be called the native's point of view is indeed subject to much debate. . . . But rest assured, realist ethnographies all claim to have located it and tamed it sufficiently so that it can be represented in the fieldwork report" (p. 50).

My own use of closely edited quotations from my interviews with Rachael begins on the third page of the article in a section called "The Participant" (1 1/2 pages). Here, using her words, I summarize Rachael's early years, her family, and her development as a sportswoman. I end this section with a dramatic key quote to highlight the tumor and its consequences as an epiphany, or major turning point, in her life. Thus, the epiphany is set up as the focal point of the article, and I devote another 7 1/2 data-based pages to dealing with Rachael's reactions to it in terms of identity dilemmas and loss of self.

In a similar fashion, extensive closely edited quotations characterize the articles by Gould et al. (1997) and Young et al. (1994). In the former study, 140 raw-data themes were extracted from interview data relative to the coping strategies used by the athletes involved. The inductive analysis conducted by Gould et al. (1997) reveals seven general dimensions: (1) driving through, (2) distracted self, (3) managed emotions/thoughts, (4) sought and used social resources, (5) avoidance and isolation, (6) took note and drew on injury lessons, and (7) other. Each of these general dimensions along with associated higher order themes are then discussed in the results section, which makes up just over 10 pages and includes two tables. The results section relies heavily on the selected quotations of the participants to provide evidence for the dimensions and higher order themes, to substantiate any claims made, and to add depth and substance to the bones of the information provided in the two tables.

In the article by Young et al. (1994), an inductive analysis yielded several themes: early influences, injury talk (pain principle and principled pain, hidden pain, disrespected pain, unwelcomed pain, and depersonalized pain), and injury adjustment (disempowered masculinity; recovered, reframed, and unreflexive masculinities). Each of these themes was supported by numerous, and often extensive, quotes from the athletes involved. In all, 10 pages of the article deal with these findings.

The net effect of using quotations in each of these articles is to give the reader a strong sense of the participants' voices. With my own article, the volume, density, and organization of the quotations as "data" suggest that Rachael is telling the story as it happened and that I just happen to be passing it on to them untouched. Of course, this is an illusion. As Van Maanen (1988) emphasizes, a great deal of typographical play, stage-setting ploys, and contextual framing goes into presenting the participant's point of view, and some tricky epistemological stunts are performed on the ethnographic high wire. Such fancy footwork is rarely discussed by researchers when they construct their realist tales.

For example, part of the "fancy footwork" I performed as the author was to take key issues and extracts from 18 hours of interviewing that was wide-ranging in content and conducted over a two-year period. Thus, quotes and events are taken out of sequence for my theoretical purposes as I invisibly orchestrate Rachael's commentary. Then, as part of my inductive analysis, I reconstruct various strands of Rachael's story into a research story that is presented in an orderly and linear fashion to the reader. In this sense, it is not so much Rachael's story that the reader gets, but a story that I have constructed about her, and my interpretation of it.

The same might be said for the articles by Gould et al. (1997) and Young et al. (1994). In each, there is a strong sense of participant voices being foregrounded. These voices operate convincingly as forms of evidence to support the themes identified and the claims made by the authors. As in my own article, however, the participant voices are nonetheless orchestrated. As Tedlock reminds us,

> Even when a few native lines do appear they seem to have been inserted for the sake of illustrating some point the author was already trying to make.... the only kind of dialogue they put out front in ethnographies is one which the natives speak briefly, on cue, and in harmony with the views of anthropologists.
>
> *Tedlock, 1987, p. 325*

Furthermore, when attempts are made to represent multiple subjects, the resultant texts tend to portray people as "flat," unidimensional, highly stable, and predictable characters, as opposed to multidimensional "rounded" characters.

Interpretive Omnipotence

The convention of interpretive omnipotence, according to Van Maanen (1988), works in several ways. Sometimes a cultural description is tied to a theoretical problem of interest to the researchers' disciplinary community. Data are then put forth as facts marshaled in accordance with the light they may shed on the generic topic and the researcher's stand on the matter.

> The interpretations of the author are made compelling by the use of a string of abstract definitions, axioms, and theorems that work logically to provide explanation. Each element of the theory is carefully illustrated by empirical data. The form is aseptic and impersonal, but it is convincing insofar as an audience is willing to grant power to the theory. . . . A realist tale offers one reading and culls its facts carefully to support that reading. Little can be discovered in such

texts that has not been put there by the fieldworker as a way of supporting a particular interpretation.

Van Maanen, 1988, pp. 51-53

Article titles, abstracts, and introductory sections that review the literature provide powerful framing devices for what is to follow and how it should be interpreted by the reader. For example, my own title signals that athletic identity can be a problem for the self. The abstract confirms this idea and signals that a story will follow about the loss of self, identity dilemmas, and the problems of narrative reconstruction for an elite performer after a major biographical disruption that terminated a promising career. The introduction to the article then proceeds to reinforce these issues by locating them in a referenced, learned, and academic literature (i.e., a selective literature review).

Golden-Biddle and Locke (1997) consider the importance of introductions and literature reviews in crafting a storyline for academic articles. They identify two key processes that, in combination, compose the article's complications and foreshadow its resolutions.

In the first process, constructing intertextual coherence, the articles re-present and organize existing knowledge. The articles variously construe the extant literature as synthesized, progressive, and noncoherent. In the second process, problematizing, these articles subvert the very literature they just constructed that provides the location for the research. They subvert it by identifying a problem variously construed as a gap, an oversight, or a misdirection.

Golden-Biddle & Locke, 1997, pp. 25-26

In the sense expressed in Golden-Biddle and Locke's quote, my own article combines the literature review and the sections on data analysis and interpretation to construct intertextual coherence by appealing to *synthesized coherence.*

Constructing synthesized coherence puts together work that previously had been considered unrelated. It highlights the need for new work (e.g., the present study) by disclosing an undeveloped investigative concern that is common to the referenced work. Studies that are otherwise unrelated are connected by constructing congruent relationships among the different referenced research streams and studies.

Golden-Biddle & Locke, 1997, p. 29

Thus, in my introduction, I call on work on the self in chronic illness, illness narratives, and the psychology and sociology of athletic identity and injury. The sections on data analysis and interpretation make explicit the three major frameworks that initially informed and shaped the study. These were the work of Charmaz (1987) on the struggles for self, preferred identities, and identity hierarchies in coping with chronic illness; the five-stage model of dramatic self-change proposed by Athens (1995); and the three narrative types called on during serious illness as identified by A. Frank (1995). The implicit suggestion throughout is that by combining these seemingly disparate strands of thinking together to focus on the case of one elite sport performer, new understandings might emerge. As such, I problematize the situation by hinting at *inadequacy.* "Illuminating oversights is the hallmark of texts that problematize the field as inadequate. These texts claim that the extant literature. . . . has overlooked perspectives relevant and important to better understanding and explaining the phenomena" (Golden-Biddle and Locke, 1997, p. 37).

Finally, it goes without saying that people like Athens, Charmaz, and Frank are leading scholars in their particular fields. By calling on them I enhance the power of my interpretation. Indeed, as Van Maanen (1988) notes, doing so allows me, the humble fieldworker, "to stand on the shoulders of giants (and see farther) by using well-received constructs as receptacles for field data" (p. 52).

Having introduced the different but powerful, prestigious, and accepted theoretical frameworks that I have used to interpret the data, and by implication the frameworks I want the reader to buy into, I then proceed to further assert my interpretive omnipotence by the skillful use of the participant's point of view as outlined in the previous section. Here, extensive quotations from Rachael are provided as "evidence" under the following section titles that constitute the bulk of the article: the emergence of a high-performance body; feelings of loss and fragmentation; the demise of the disciplined body-self; the demise of the gloried self; and finally, holding on to past selves. In each, the quotations support the theoretical concepts put forward by myself (read Athens, Charmaz, Frank, and others) to explain what is happening in Rachael's life.

By resting my case on what Rachael says and does, I call on another device to enhance interpretive credibility. The data are presented conventionally as the events of everyday life. These situations, along with a generalized rendition of Rachael's point of view, are collapsed into explanatory constructs in such a way that my analysis overlaps with (if it does not become identical to) the terms and constructs used to describe the events. Here, as Van Maanen (1988) comments, the author's authority is embellished by using the "native's" vernacular to suggest that the author is fully able to whistle native tunes.

At the end of my article come two sections that are designed to narrow the range of interpretations on offer. The first section, called "Narrative Constraints in Reconstructing the Self," is an overt return to "pure" theory. No words of Rachael are here. Her words from previous sections are explained, however, by referring directly to the concepts and key scholars mentioned at the start of the article. Finally, my interpretation of events—that a strong athletic identity can act as an Achilles' heel when an elite performer experiences a disruptive life event, and that a narrative form of analysis is useful in understanding this phenomenon—is solidified. Having funneled the reader to this conclusion, I close by considering the implications of my "findings" for health care professionals.

The articles by Gould et al. (1997) and Young et al. (1994) follow lines similar to my own. The title and abstract frame the article by Gould et al. as being within a specific and recognizable tradition in psychology rather than another discipline. Next, the literature on injury is reviewed in relation to the related topics of athletic injury, stress, and coping strategies. Here, Gould et al. construct intertextual coherence by appealing to *progressive coherence*. According to Golden-Biddle and Locke (1997), this strategy "incorporates works already recognized as related in theoretical perspectives and methods" (p. 31). Next, Gould et al. proceed to problematize the situation via an appeal to what Golden-Biddle and Locke call *incompleteness*. "When problematizing the literature as incomplete, the article claims the extant literature is not fully finished; it seeks to contribute by specifying and filling in what is not finished" (p. 36).

Commenting on the available literature, Gould et al. include remarks like, "The lack of attention paid to the role that psychological factors play in athletic injury rehabilitation and recovery is somewhat surprising" (1997, p. 380); "not only is little known about coping strategies used or their effectiveness in injured athletes, little is known about whether males and females use similar coping strategies when injured" (p. 380); and "no studies have been conducted to specifically assess the types of coping strategies used by injured athletes" (p. 381). Having identified the gaps in the literature, Gould et al. make the case that "a need exists to identify and examine coping strategies used by injured athletes. This study addresses this need" (p. 380).

Having set up the need for their own study, the authors proceed to outline the research on coping strategies that informs their work. The methods and the results section considers how the study was conducted, and interview data are presented as evidence of the perspectives and themes identified in their hierarchical content analysis. The closing discussion is framed by the article's specific purposes. The study's findings are often linked to the findings of other leading scholars in supportive and confirming ways. The importance of the findings from a practical perspective are also highlighted before suggestions are made for future research.

The article by Young et al. (1994) follows a similar format. The introduction and the following section on "physical power and symbolic masculinity" identify the key sociological concepts and theoretical frames that guide the data presentation and analysis of the "body talk" of male athletes to show "how men learn to disregard risk of physical harm and to normalize pain and disablement as part of the sport experience" (p. 176). As such, the introduction constructs intertextual coherence via progressive coherence. The summary and discussion at the end of the article pull its various stands together in a confirmatory manner as the authors note, "What is learned through 'injury talk' is the significance of physicality for the hegemonizing of masculinity based on notions of dominance" (p. 192).

Finally, the conclusion and discussion sections of each of these articles play their part in nailing down certain interpretations.

> In brief, the ethnographer has the final word on how the culture is interpreted and presented. . . . rare are ethnographers who question aloud (or in print) whether they got it right, or whether there might be yet another, equally useful way to study, characterize, display, read, or otherwise understand the accumulated field materials. . . . Self-reflection and doubt are hardly central matters in realist tales.
>
> *Van Maanen, 1988, p. 51*

Therefore, it would seem that my own article and those of Gould et al. (1997) and Young et al. (1994) adhere closely to the central conventions of realist tales and gain their persuasiveness from these conventions. These are summarized by Cole as follows:

> The account is given increased legitimacy through representations of actual subject voices selected and excluded based on their consistency with the author's "report." While a sense of authenticity and objectivity permeate the narrative, the ethnographer translates the voices and visions of local subjects into his/her own. The analysis (or the view from above) is bounded and legitimated by monitoring incongruent voices or anomalous moments; thus, fixing and stabilizing theory and limiting alternative interpretations by readers.
>
> *Cole, 1991, p. 40*

Having identified how conventions work in realist tales, I must acknowledge these conventions are neither good nor bad in themselves—a point I return to at the end of the chapter. Nor does it mean there is no room for maneuver within the conventions identified. That is, modifications are possible within certain limits.

Modifying Realist Tales

The realist tale tends to be author-evacuated. At times, however, for certain purposes, it may be both possible and desirable to modify this tale. One way, although it might seem rather mechanistic, is for authors to indicate their social positioning. This has been done by Hanson in a 1992 article titled "The Mental Aspects of Hitting in Baseball: A Case Study of Hank Aaron" and by Newman in his 1992 article titled "Perspectives on the Psychological Dimension of Goalkeeping: Case Studies of Two Exceptional Performers in Soccer." Each author includes in the methodology part of his article a section titled "Instrumentation: Person as Instrument Statement." Here, each gives his current employment status, appropriate qualifications, theoretical interests, and biases with regard to the topic of his analysis. While this information on the author is useful, it is essentially descriptive and it acts as an additional means to persuade the reader of the author's *credibility* (see the earlier discussion on experiential authority). As Newman comments, "To establish the researcher's credibility as an effective instrument of inquiry, it is necessary to provide information regarding his background, qualifications and training" (p. 77). It is also interesting to note how Hanson talks in the first person, as "I," in this section, whereas Newman prefers to talk about himself in the third person, as "the researcher" and "he."

In their articles, both authors provide details of the social class origins of the participants in their case studies. Hanson (1992) also identifies the ethnicity of his subject by providing, for example, details of the death threats that Henry Aaron (a Major League baseball hitter) received "from people not wanting a black man to break the legendary record [of Babe Ruth's career home run records]" (p. 51). In both articles, however, we are left to ponder the ethnicity of the authors and their social class origins, along with the possible impact, if any, that these had on the interactions that took place as part of both the interviewing process and the production of the final written account. Finally, Hanson announces his presence in the text at various points as a named voice in verbatim transcriptions (presumably selective) of an interview between himself and Hank Aaron. In contrast, Newman (1992) is absent from his presentation of salient themes that emerged in his interviews with two international goalkeepers—Pat Jennings of Northern Ireland and Tony Meola of the United States. In essence, while both authors provide some details of their positioning and acknowledge their authorial presence in specific sections of the text, they essentially rely on realist forms of representation to tell their stories.

I have also constructed a modified realist tale to highlight the experiences of oppression that a lesbian physical education teacher and sportswoman, Jessica (a pseudonym), experienced in her daily working life in a secondary school in England (Sparkes, 1994b). Given the power differentials that operated in our emerging relationship and the production of the final written text, I felt it essential to name, in a separate section of the article, the social categories that I belonged to in relation to Jessica. Despite our commitment to a collaborative relationship, this step was necessary to signal not only the privileges that came with my membership of such categories as male and heterosexual but also the ways in which these permeated my interpretation and the construction of the final text in relation to Jessica as female and lesbian. As I constructed this account, I was acutely aware of the issues relating to voices in the text. After much thought and several discussions with Jessica, I opted for a strategy that included theory at the beginning and end of the article, with the middle section focusing on selected moments from Jessica's life history. This middle section relied heavily on Jessica's own words to illuminate her experiences of homophobia and heterosexism in different contexts and the manner in which these effectively served to silence her and render her invisible in specific social situations. Therefore, I only announce my authorial presence at various points in the text and become an "absent presence" once Jessica's voice is introduced.

Of course, my "disappearance" is itself a textual illusion because I am ever present throughout the article as its author, and it is my guiding hand that selects the quotations and shapes the story presented. Therefore, my disappearance needs to be seen as a textual strategy, a conscious decision, to focus attention on Jessica's words with a view to drawing the reader into the storyline of oppression and evoking a response. Essentially, I wanted readers to feel Jessica's oppression and begin to locate themselves in the dynamics of this process. I wanted to evoke a response to Jessica's situation by providing what Hooks (1991) calls a *narrative of struggle,* a narrative in which the subjectivity of the oppressed individual reasserts itself. Whether or not this tactic works only the reader can tell. It does, however, signal that realist tales can be used to serve a critical agenda. By describing how the life-worlds of people who appear as strangers are shaped by particular webs of contingencies within a sociopolitical milieu, the author hopes that the reader can achieve solidarity with those described as fellow human beings (see Barone, 2000).

Realist tales can also be modified to include different narrative styles and conventions. This may be easier to accomplish in books and monographs rather than in articles in journals, since in the former the author often has more space in which to experiment and perhaps greater freedom of expression away from the harsh eyes and pens of referees. Here, a good example is provided by Klein (1993) in his book *Little Big Men: Bodybuilding Subculture and Gender Construction*, which is based on a seven-year ethnographic study of four elite gyms on the West coast of the United States.

These four gyms are amalgamated (in an analytical sense) to create one fictitious gym called *Olympic Gym*, which is the focus of the book. Despite this fictive element, *Little Big Men* is essentially a realist tale in that the majority of the chapters draw on the conventions of this genre as described earlier. For example, the author is absent from most of the text, but, as Klein acknowledges in an appendix, the early chapters reflect his attempt to experiment with different forms of representation to situate the reader in the world of competitive bodybuilding. To this end, he felt it important to distance and to draw in the reader simultaneously, because distancing can help to suspend a priori knowledge of the subject while bringing the reader closer can facilitate translation. Therefore, to negate or defamiliarize the reader's previously held notions about bodybuilding, Klein exaggerates certain aspects of this subculture by representing it in a comic vein that highlights the ludicrous and carnival aspects of gym life.

To combat the fact that some dimensions of the gym scene were so similar to commonplace urban North America, Klein made use of powerful metaphors from the general culture along with the technique of *diorama* or *microcosm* to create the distance required. For example, in the following passage, the gym is portrayed in terms of a city.

> Viewed from the balcony of the 10,000-square-foot gym, one can see that it is laid out gridlike, as if composed by Salt Lake City's Mormon founders. Main street: a neat avenue about six feet wide and 250 feet long bisects the gym. Weight stations, miniskyscraper affairs of iron in a variety of shapes and functions, cut the space into cubes and line the main boulevard the entire length of the gym. "Avenue of the Olympians," I call it. Jutting off at regular intervals are straight little streets variously leading to pulleys, back machines, free weight racks, and the like—all dead-ending into mirrored walls that create an infinite sense of space. There's Leg Press Lane, Bicep Boulevard, and Ab Avenue. There is even a ghetto of "Blue Monsters" (Nautilus machines). The front desk, located at one end of the main avenue, magically becomes the old town hall where all official business gets transacted. . . . At times the streets of Olympic Gym are virtually deserted, and only a few inhabitants are about. However, around noon

and six in the evening—rush hours—the gym is crowded with bodies double- and triple-parked: big, eight cylinder jobs with gleaming chrome deltoids and baroque hood ornaments that double as chests.

Klein, 1993, p. 22

In looking at the gym as a city, Klein (1993) attempts to magnify it, and so distance it, in the same way that a camera pulls back on a tightly focused shot to reveal a larger subject. Likewise, in discussing his description of a typical night at the gym, Klein notes how he employs a slightly carnivalesque characterization of life at Olympic Gym that is both a literary and an accurate depiction: "This has the effect of creating a tension between known and unknown elements of gym life; the commonplace knowledge is juxtaposed against the slightly ludicrous in the hope of situating the reader and providing a heretofore unencountered ambiance" (p. 289). In addition, the opening chapters also attempt to defamiliarize by utilizing a reflexive style that decenters the reader by drawing attention away from the ethnographer-as-authority. For example, Klein periodically invokes an autobiographical voice with the intention of drawing himself-as-ethnographer and the reader "outside the objective perimeter of the study by encouraging self-reflection by both parties. In this way interpretation becomes a three-way interaction between ethnographer, reader, and the subject(s) of the study" (p. 289).

As the work of the scholars discussed in this section illustrates, even though there are a set of key conventions that frame realist tales, this frame is not rigid or impermeable. As a consequence, those who feel increasingly uncomfortable about producing author-evacuated tales might consider writing more of themselves into the text when, for certain purposes, they feel this to be appropriate. Likewise, these authors might also consider acknowledging and sharing with readers, in a reflexive manner, the rhetorical devices they have called on. Such inclusions would certainly not detract from the power and persuasiveness of realist tales and could act to enhance their ability to provide insights into the world of sport and physical activity.

 ## Reflections

By focusing on realist tales of athletic injury as told by Gould et al. (1997), Young et al. (1994), and myself (Sparkes, 1998a), I have indicated the conventions associated with this genre. The participants' voices are foregrounded in these tales, and the reader is able to gain important insights into the participants' perceptions of the injury experience. The orchestration and theoretical framing of these voices by a disembodied author, however, are conventions that resonate with certain conventions of the scientific tale. This similarity worries some researchers, but it does not make the conventions themselves either good or bad. The realist conventions connect theory to data in a way that creates spaces for participant voices to be heard in a coherent text, and with specific points in mind. When well constructed, data-rich realist tales can provide compelling, detailed, and complex depictions of a social world.

In terms of the ongoing development of realist tales, A. Fine (1999) makes a strong case for a reinvigorated realist ethnography. The core justifications of this approach include the collection and presentation of extensive, high-quality, rich, and persuasive descriptive data. He suggests that "realist ethnographers do not and should not maintain that the descriptions they present are exact or transparent accounts of a world out there" (p. 535). Noting that all systematic studies have limitations, he goes on to warn against ontological despair. That is, ethnographers do not need to reach a precise, definitive, singular truth in order to have something useful and important to say about the contemporary human landscape. Accordingly, A. Fine offers three justifications for realist tales. These are the intersubjective, epistemological, and pragmatic defenses.

With regard to the intersubjective defense of realist tales, A. Fine (1999) notes that even if a set of truth claims cannot be said to be definitive, they can typically serve us well enough. He also acknowledges that if two ethnographers visited the same scene, they might leave with two very different pictures. However, as A. Fine is quick to point out, the pictures that they perceived would be recognizable to each other. They have intersubjective power, and on this basis readers of realist tales will draw their conclusions as to whether they are "true" or not.

> One of the features that has made classic realist ethnographies so powerful is that they make sense to readers and to participants of the scenes described. As an audience, we can picture these scenes, and these scenes accord with what we know of those settings.
>
> *A. Fine, 1999, p. 536*

Regarding the epistemological defense, A. Fine (1999) notes that the legitimacy of realist ethnography is based on the recognition that the way researchers come to learn about the world is precisely the way in which the participants in the social setting come to learn about their world, a world that they tend to believe has an obdurate quality. In coming to know in the same manner as participants, the researcher serves as a well-informed and, at times, expert guide to scenes and social worlds that the reader may not have encountered.

Finally, in terms of a pragmatic defense, A. Fine (1999) notes that realist tales are surely *useful*. For him, realist accounts permit us to build knowledge, comfortable in the belief that we are learning about the contours of the world. A. Fine acknowledges that the question of truth may be unknowable on a deep, epistemological level. But, he adds, it is possible to judge knowledge by its results: "There can be little doubt, even among the critics of the realist approach, that some of the classic naturalistic ethnographies have produced startling, compelling, and practical results that have demonstrated their worth" (p. 538). Therefore, realist tales do have a contribution to make. As Van Maanen (1988) comments:

> It is important not to judge realist tales too harshly. Realist ethnography has a long and by-and-large worthy pedigree. Writers in this tradition have created masterpieces that have lived very long lives. To subject the writing to scrutiny is not to say it is false or wrong.
>
> *Van Maanen, 1988, p. 54*

Even though the majority of us will not create any masterpieces, I see no reason why realist tales should not continue to make a significant contribution to our understanding of sport and physical activity. I certainly feel they are a useful part of my research repertoire (see Dévis & Sparkes, 1999; Faulkner & Sparkes, 1999; Holt & Sparkes, 2001; Sparkes & Smith, 1999, 2002; Squires & Sparkes, 1996). Indeed, I will endeavor to become more skilled in the use of this genre and a better teller of this kind of tale. This can only happen if I can understand the conventions of this tale so as to bring them within my methodological awareness. Having this kind of awareness encourages me to reflect on when, and why, I represent my research findings in a realist fashion. This awareness also encourages me to experiment and play with other ways of telling about the same phenomenon. These alternative tales will be considered in the chapters that follow.

Chapter 4

Confessional Tales

In realist tales, the voice of the researcher, if heard at all, tends to be
found in the methods section of the
article. Here, there may be hints that
the author is concerned about various
issues associated with the fieldwork,
such as the ethics of different relationships as they develop between the researcher and the participants in the
course of an inquiry. Such concerns,
however, are not dealt with in any great
depth in realist tales. In contrast, there
is a tale, the *confessional*, that foregrounds the voice and concerns of the
researcher in a way that takes us behind the scenes of the "cleaned up"
methodological discussions so often
provided in realist tales.

According to Van Maanen (1988),
the fieldwork confessional is a representational genre that contrasts sharply
in a number of ways to the realist tale.
For example, realist tales tend to be
author-evacuated and methodologically silent; they adopt the conceit
that data must be cleanly separated
from the fieldworker, and they offer
only the researcher's tightly packaged
and focused account of the culture
studied. In contrast, the two main

distinguishing features of confessional tales are their highly personalized styles and their self-absorbed mandates. According to Smyth and Shacklock, these characteristics fill the space in research that is left by the phenomenon of the "missing researcher."

> The reflexive narratives of researcher's encounters with the inter-sections between the researcher's values and the research processes reintroduces the researcher as person into the account. Issues like: ethics, gender, race, validity, reciprocity, sexuality, voice, empower-ment, authorship, and readership can be brought into the open and allowed to breathe as important research matters.
>
> *Smyth & Shacklock, 1998, p. 1*

Confessionals, therefore, explicitly problematize and demystify field-work or participant observation by revealing what actually happened in the research process from start to finish. Therefore, the details that matter in confessional tales are those that constitute the field experience of the author. These include, according to Van Maanen (1988), "Stories of infil-tration, fables of fieldwork rapport, mini dramas of hardships endured (and overcome), and accounts of what fieldwork did to the fieldworker" (p. 73).

For example, Boman and Jevne (2000) draw on their actual experience of making a "mistake," which involved being charged with an ethical vio-lation for disclosing the identity of a study participant. The violation is presented first as a narrative account from the perspective of the researcher involved (Boman); next, the authors move on to consider different ways in which the situation might be judged. Consequently, an experience that might initially be viewed as a researcher's nightmare is redefined as a useful resource for raising questions about ethical practice in qualitative research, and about how these questions might be answered by research-ers learning collectively from the difficulties and problems that they have encountered individually in the field. In effect, what Boman and Jevne tell is a confessional tale about the dilemmas and conflicts of the research ex-perience that are more often than not left untold.

Even though unmasking fieldwork is a relatively recent phenomenon, Van Maanen (1988) feels the genre has become a fairly large one. Indeed, he argues that providing a confessional to supplement realist reports of fieldwork has become a more or less institutionalized feature in both soci-ology and anthropology: "It is pro forma these days to append a confes-sional to a fieldwork dissertation or to include one in a separate chapter of the thesis under the 'methods' label" (p. 81). In contrast, Smyth and Shacklock (1998) note that it is often difficult to find accounts of the com-plexities of doing critical research, written by researchers who construct the accounts out of critical appraisals of their own work. As a consequence, they see a need for deeply personal and individual readings of "the experience of critical research" in educational and social settings.

> We saw the need for individual researchers to present accounts of the experience of the critical research act using reflective postures that challenged why one course of action was taken among a range of possibilities. Such accounts . . . should focus upon those issues and dilemmas which caused trouble and uncertainty in the research process. As we saw it, such accounts would tell a story about the intersection of the critical research perspective and the particular circumstances of the research context.
>
> *Smyth & Shacklock, 1998, p. 1*

With these points in mind, in this chapter I have chosen to focus on articles from sport and physical activity that I define as confessional in character. That is, they draw on personal experience with the explicit intention of exploring methodological and ethical issues as encountered in the research process. This definition is necessary both to understand my selection and to distinguish these confessional tales from the autoethnographies discussed in chapter 5. Although each draws on autobiographical material, the confessional exists in a symbiotic relationship to realist tales, whereas autoethnographies are a distinct genre that stand by themselves. Confessionals are also distinguished by other conventions that will now be considered.

Confessional Conventions

Confessional tales appeal to *personalized author(ity)* and emphasize the *researcher's point of view*. As Van Maanen (1988) notes, "Author-fieldworkers are always close at hand in confessional tales" (p. 74). Such writing is intended to show how each particular work came into being and to reveal the dilemmas and tensions contained in the process. The fieldwork odyssey, and the process and problems of coming to know, are the main focus, rather than the findings. Thus, the ubiquitous, disembodied voice of the realist tale is replaced by the personal voice of the author, announcing, "Here I am. This happened to me and this is how I felt, reacted, and coped. Walk in my shoes for a while." This is an important move as there is an intimacy to be established with the readers, personal characters to develop, and trials to portray.

Accordingly, the human qualities of the researcher are often foregrounded, and an acknowledgment of personal biases, character flaws, anxieties, and vulnerabilities helps to develop an ironic self-portrait that readers can identify with. The message is: "I'm just like you. I get hurt, get it wrong, get misled, and struggle to understand." Therefore, as Van Maanen (1988) points out, "The omnipotent tone of realism gives way to the modest, unassuming style of one struggling to piece together something reasonable and coherent out of displays of initial disorder, doubt, and difficulty" (p. 75).

Another feature of confessional tales is that the author is rarely portrayed as a passive actor who simply waits for things to happen. More often, the author is an active agent, usually working in difficult circumstances. As Van Maanen notes:

> The ethnographer as a visible actor in the confessional tale is often something of a trickster or fixer, wise to the ways of the world, appreciative of human vanity, necessarily wary, and therefore inventive at getting by and winning little victories over the hassles of life in the research setting.
>
> *Van Maanen, 1988, p. 76*

Furthermore, Van Maanen (1988) adds, fieldworkers in the confessional tend to be blunt about the ways in which ethical principles often have to be compromised to get the job done. Such portraits may not be flattering, but they can be brutally honest about how research often has to take place.

The fieldworker's point of view is often represented in confessional tales as part of a character-building conversion tale in which the researcher, who had a view of how things might happen at the start of the study, comes to see things very differently as the study progresses. As part of this process of coming to know and, by definition, getting closer to the participants' view of their world, confessional tales often include episodes of fieldworker shock and surprise. The blunders and mistakes made, the social gaffes committed, and the secrets unwittingly unearthed are also revealed as part of this coming to know.

The final convention of confessional tales identified by Van Maanen (1988) is that of *naturalness*. This relates to the manner in which, despite all the bothersome problems and dilemmas revealed in the confessional tale, it nearly always ends up supporting whatever realist writing the author has published previously.

> Though confessional writers are forthcoming with accounts of errors, misgivings, limiting research roles, and even misperceptions, they are unlikely to come to the conclusion that they have been misled dramatically, that they got it wrong, or that they have otherwise presented falsehoods to a trusting audience. The implied story line

of many a confessional tale is that of the fieldworker and the culture finding each other and, despite some initial spats and misunderstandings, in the end, making a match.

Van Maanen, 1988, p. 79

Therefore, confessional tales tend to exist in a symbiotic relationship to the realist tales told about the same research. As such, they do not replace realist accounts but stand beside them by elaborating extensively on the formal elements of the methodology and saying what is unsayable in the realist telling. In view of this feature, as P. Atkinson (1991) suggests, perhaps the confessional tale is not really an alternative genre, because it exists in a complementary rather than contrasting relation with realist tales.

These stylistic conventions form a backdrop to the confessional tales that have begun to emerge in sport and physical activity. The exemplars in the next section call on each of these conventions to a greater or lesser degree. They also signal a growing acceptance of this tale in sport and physical activity.

Confessionals in Sport and Physical Activity

In his book *Little Big Men*, Klein (1993) incorporates a confessional element in a realist tale. In the opening chapter is a section titled "Confessions of an Anthropologist." Here, Klein reveals some moments of embarrassment and confusion he encountered as he tried to make sense of life in the gym. For example, even though he had prepared himself to observe the subculture of bodybuilders, Klein admits to being caught off guard and reacting only to the intimidating size of these people when he entered the gym initially. He coped by giving his attention to the pictures of bodybuilders (mostly male) posing on the wall.

> I was trained in techniques of observation. . . . I was convinced that there was not a bizarre or grotesque type of behavior I hadn't seen, read about, or had told to me. Crossing the threshold of the gym door, however, I unexpectedly froze when it came to engaging the "erotic" scene before me. I turned instead and frantically examined the anonymous wall behind the front desk. The embarrassment I felt over watching the goings-on was bad enough; my response was something else—very unprofessional. . . . "Can I help you?" The voice of someone behind the desk filtered through my concealed terror. I turned, frantically trying to think of something to say, some reason for being there, for standing like an idiot looking at these pictures. How embarrassing, had I been gawking for an eternity?

> *Klein, 1993, pp. 24-25*

Besides moments of embarrassment and confusion, Klein (1993) also gives details of how, at times, the bodybuilders knowingly misled him in terms of the information they provided. One incident, in particular, forced Klein to acknowledge his naiveté. It involved a recognized bodybuilder who informed Klein that, although he had at one time taken steroids, he no longer used them because he had mastered the intricacies of diet and training. Two days later Klein heard the same bodybuilder joking with his colleagues about an advice column he wrote in a leading bodybuilder magazine and the views he had expressed in response to a young bodybuilder's inquiry about steroid use. The comments made, the laughter, and the jokes about what he had said in the advice column indicated that this bodybuilder was still taking steroids. Such moments served as indicators of a ritual of passage into the bodybuilding culture that led Klein to conclude that "being gullible was, oddly, an integral part of the field experience. Initial data posed a threat if one took it at face value, and the measure of cultural understanding came in direct proportion to the ability to discern and play with (interpret) behavioral contradictions" (p. 29).

Feminist and profeminist scholars, in particular, have utilized confessional elements to reflect on, examine critically, and explore analytically the ethical and methodological concerns embedded in their work (see Clarke & Humberstone, 1997; Hall, 1996). A reflexive position is seen as important for helping researchers to locate themselves in the power dynamics of the research relationships (such as researcher/researched or researcher/funder) and for developing a healthy skepticism toward their own findings (see chapter 1 of this book). For example, my own confessional tale (Sparkes, 1994a), published in the same year as my realist tale of the same topic (Sparkes, 1994b), focuses on the emerging relationship between myself, a white, heterosexual male, and "Jessica," a white, lesbian physical educator and sportswoman who became involved with me in a life history study.

My experiences in this relationship highlight that the collaborative movement from life *story* to life *history* is problematic and laden with ethical and methodological dilemmas, as well as transformative possibilities. These dilemmas are discussed in relation to issues of biographical positioning with regard to difference and sameness, sharing stories and building trust, the nature of collaboration, the researcher as therapist, friendship, and the postmodern concern over representation and authorship. In contrast to the realist tale, which is for the most part author-evacuated, my confessional tale reveals a great deal about me as a person and about how my background shaped my understanding of Jessica's experiences. It also reveals how I came to change my thinking over time as I grappled with key ethical dilemmas. For example, in discussing the impact of listening to Jessica's stories, I note how I became increasingly aware of my actual involvement in them as a researcher:

> As our interviews progressed, I became increasingly aware of my own tacit involvement and entanglement in the strands of homophobia that continue to shape aspects of Jessica's life, and mine. . . . For example, I became aware that although I taught an undergraduate module that focused critically on a number of ideologies that pervade PE, neither homophobia nor heterosexism were included. My involvement in Jessica's story highlighted this as a significant silence on my part and led me to reshape this course in order to explore how these ideologies act to oppress specific groups of people.
>
> *Sparkes, 1994a, pp. 175-176*

Aspects of my confessional tale are extended in another tale (Sparkes, 1998b), where I revisit my seven-year relationship with Jessica to raise critical questions about the nature of reciprocity in our relationship. This looking back enables me to see all of the old complexities in a much more scoped fashion; for example, the motives of self-advancement in constructing research accounts, the skewed nature of the benefits, and my own continual need to reaffirm and renegotiate the nature of the tensions I experienced in the original research process. In this looking back, which involved further collaboration with Jessica, there is a strong element of "unsettling" around the multiple meanings of reciprocity in critically informed life history research.

Humberstone (1997) also draws on her personal experiences to consider the tensions involved in an ethnographic research project on physical education and outdoor education in which she operated as a participant observer for three months at an outdoor education center. The reflexive account makes visible the ways in which she, as the researcher, influenced the investigative process even as she was influenced by it. In particular, her confessional tale highlights the ways in which she had to navigate through webs of power, and the tensions and conflicts this initiated as she attempted to remain true to her feminist principles in relation to her analysis and the criteria used to judge her work, which was submitted for a PhD. Humberstone's account does not try to hide these problems but rather explores and exposes them.

In a similar fashion, Pedersen (1998) provides a reflexive account of her experiences of conducting an ethnographic study for her PhD in a "wilderness" community located in far northern Norway. The aim of the study was to examine how the ideologies, practices, and experiences of outdoor life structured relationships between women and men and how this, in return, affected ideologies of gender. In her article, Pedersen highlights how a number of different methodological concerns arose during the course of her fieldwork with regard to relationships between her as the researcher and the participants as gendered actors, and between her and the social and cultural contexts of the research process. In particular, her personal experiences of a web of multiple power relations highlight a number of dilemmas Pedersen had to cope with in women-only, men-only, and gender-mixed but male-dominated settings.

For example, Pedersen (1998) notes that as a female researcher, in female-only sites she tended to be regarded as an experienced "outdoor life person," a leader, and a teacher. In gender-mixed and in men-only field sites, however, she was rarely given, or able to take on, such a role; her role depended on the activity in question. Thus, Pedersen was treated more equally by men in activities that did not require specific knowledge, skills, equipment, or technology, or that were not perceived as important symbols of masculinity, and among men who did not need the wilderness to confirm their masculine identity. Pedersen did not participate in many men-only field sites, however, because she found it difficult to invite herself on these "buddy-trips," and she was rarely invited spontaneously. When she was, she felt uncomfortable with the sexual jokes that the men told in these settings and with their heroic stories of mastering the wilderness, which made Pedersen and other women listen in silence. Despite her discomfort as a feminist in adopting these roles, Pedersen acknowledges that she had to compromise her principles to access the local male culture. These principles are tested at other points in her fieldwork as Pedersen is subjected to sex stereotyping and sexual harassment.

> Even more uncomfortable experiences occurred in relation to some of the male informant who were older than me: one hugged me around the waist from behind and another started to "foot-flirt" during an informal conversation. I tried to avoid these approaches in a friendly manner as I did not want to lose information. . . . Not all interactions with male informants were like this. On some occasions I was treated like a daughter or as a possible spokesperson for their interests, perhaps because I was conscious of. . . . taking the role of the novice. . . . Frequently it reflects upon how a female researcher's fragile identity can be threatened by being treated as a sex object, a daughter to be protected, a servant, a mascot, or a non-professional.
>
> *Pedersen, 1998, pp. 398-399*

Pedersen (1998) draws on such experiences to ask questions about reflexivity and situated knowledge, about how as gendered beings we come to know what we know, and about whose knowledge is made visible and invisible in ethnographic research. Her experiences are also useful for asking questions about what happens when the boundaries between the life of the female researcher and the field site disappear.

Boundary maintenance and survival are central to the 1999 article by Brackenridge titled "Managing Myself: Investigator Survival in Sensitive Research." Here, Brackenridge offers some personal evaluations of the challenges she faced during her 13 years of investigative research on the sensitive and controversial topic of sexual abuse in sport. Brackenridge notes how, in presenting her findings previously to broadly scientific or to practitioner groups, she has typically adopted a combination of two context-free authorial voices. One is the disembodied, pseudoscientific, positivist authority, where she presents "data," "theoretical models," and "explanations" of "risk factors" and the dynamics of sexual abuse in sport. The other voice, used in her realist tales, claims experiential authority by drawing heavily on quotations from research participants to illustrate the social processes and personal consequences of abuse. As Brackenridge comments, "In both types of writing, I am absent-as-person; my life experience as female, lesbian, white, middle-class, political activist and advocate is missing. Yet all these elements of my self-presentation must influence my research participants and my results" (p. 401).

In a real sense, as Brackenridge (1999) acknowledges, in these forms she has presented what M. Fine (1994) calls the "clean edges" of the narratives given to her as a researcher, and these contrast starkly with the "frayed borders" given in the testimonies of her research participants. These clean-edged presentations also do not reveal the frayed borders experienced by Brackenridge in conducting her work. Therefore, the primary purpose of Brackenridge's confessional tale is not to address the findings of her research but to reflect on the subjectivity of a lesbian engaged in a gendered research process.

> In particular, the intention is to use this research experience to explore strategies for personal survival as an investigator and to propose a framework for self management that other researchers might also find helpful. . . . The stimulus for this paper, which is a reflexive account of a white, middle-class, lesbian engaged in sensitive research about (largely) female oppression in a (largely) male-dominated world, was twofold: first, a legitimation crisis brought about by serial failure with grant applications; and second, an increasing burnout from the many reverses I have experienced along the way. Confronting these issues led me to articulate the previously taken-for-granted rationales for my research and recognize more clearly the social, political and historical contingencies of the work.
>
> *Brackenridge, 1999, pp. 399-400*

The confessional produced by Brackenridge (1999) is both courageous and harrowing. It reveals the pressures that researchers can be placed under when they seek to reveal, understand, and challenge what many—particularly powerful sporting organizations and academic colleagues—would prefer not even to acknowledge the existence of. In her tale, Brackenridge highlights the difficulties of managing herself in the face of stressors that include personal insults and attempted blackmail from an international coach; a threat of legal action from a national sport organization; attempted recruitment into causes by her participants; hate mail and crank phone calls following a television appearance; media harassment and misrepresentation by journalists wanting access to her data on individuals; isolation and ridicule by individuals and agencies about whom she had incriminating evidence; rejection of grant submissions on the grounds that sport research had nothing new to say about sexual abuse; and withdrawals of access by a major sport organization to an elite athlete sample for fear of what might be uncovered. To these must be added the indirect trauma and general emotional contagion of listening to, and reading about, dozens of cases of sexual exploitation, coupled with feelings of guilt and blame that she could, or should, have done more to prevent these happening.

The consequences of these combined stresses led Brackenridge (1999) to experience anxiety, insomnia, political frustration and ineffectiveness, funding shortages, and lack of primary data. There were also publishing delays and rejections based on her work being judged as either too personal or too positivistic. Such rejections are a major source of stress in an academic world where the gold standard of worth is publications in refereed journals. Not surprisingly, Brackenridge also experienced a number of personal legitimation crises during the course of her study.

In summarizing the situation, Brackenridge (1999) states, "Researching sexual abuse in sport has exacted a toll on me as an individual. . . . the messy conjunction of personal conviction and political resistance has proved, at times, almost intolerable" (p. 403). From this starting point, Brackenridge proceeds to call on additional personal experiences to explore her strategies for personal survival in relation to managing by herself—and managing her different selves (personal, political, and scientific)—in the research process. These experiences and strategies (expanded on in Brackenridge, 2001) provide a valuable resource for any researcher who is considering investigating a sensitive issue.

The confessional tale told by Sugden and Tomlinson (1999), while not profeminist, also highlights the differential power relationships between researchers and funding agencies, along with other methodological concerns. In an article titled "Digging the Dirt and Staying Clean," the authors draw on their work on the global politics of world football, related studies of other sports, and methodological issues to advocate an investigative mode of sociological inquiry, and also to warn against a reliance on standpoint epistemologies. In making their case, they cite their unique fieldwork experiences researching the Fédération Internationale de Football Association (FIFA), the governing body of world football, particularly in South Africa and France.

The confessional elements of the article are signaled in its opening story, "Touts on Tour." It tells of the wheeling and dealing, necessary deceptions, and pieces of luck required for one of the authors to gain access to Cockney Joe and other members of Fat Tony's ticket tout gang at the soccer world cup in France. Sugden and Tomlinson comment that "Getting at deep, insider information in the manner . . . is not usually advocated in graduate school research methods courses or manuals of acceptable research practice. We think it should be" (p. 386). The authors then proceed to reflect on their own particular approach to research and to ask what, if anything, is distinctive about it.

At various points, they provide confessional material to add weight to their claims. For example, in making the case for the use of "gonzo journalism" pioneered by Hunter S. Thompson, Sugden and Tomlinson (1999) note that Thompson's approach, and therefore their own (though not necessarily to such a degree), is about risk-taking, both individually and as a team. Indeed, Sugden and Tomlinson claim that the fieldwork strategy for their own FIFA project, which involved researching the structures, values, and ideologies of the governing body, "emerged from a successful gamble which one of us made while attending the African Cup of Nations in South Africa in 1996" (p. 391). The article then describes how one of the authors, having turned up uninvited, took a number of gambles, passed himself off as a journalist, and eventually got a media accreditation pass that allowed him unrestricted access to all games and personnel associated with the competition. The authors comment, "A few risks and a few white lies and we had a bus pass which would take us, via the Asian Championships, Euro 96, the European Champions League Final and World Cup France 98, into the heart of the FIFA family" (p. 391).

In a similar fashion, Sugden and Tomlinson (1999) incorporate confessional elements to support their case for developing a realist investigative paradigm dedicated to the "objective, rigorous and systematic quest for truth" (p. 393), at the expense of relativistic standpoint epistemologies. Sugden and Tomlinson make it clear that they are not against the empowerment of the marginalized and oppressed through the authentic reporting of local perspectives, but feel that if these voices are to make a contribution to "value free understanding," then these perspectives "must themselves be reevaluated through the meta-language of social science"

(p. 393). In particular, Sugden and Tomlinson are "definitely against the uncritical regurgitation of the self-serving rhetoric of the already powerful, whose charlatanism can best be exposed though the objective pursuit of truth" (p. 393).

To demonstrate their support for these principles and to illustrate what they see as the practical or operational flaws of employing a standpoint epistemology, Sugden and Tomlinson once again provide data from their own fieldwork experiences. The story begins at a chic garden party on the lawns of the British embassy in Paris two days before the World Cup Final between France and Brazil in 1998. The authors have turned up at the garden party to profile the English bid to stage the 2006 World Cup Finals. During the dinner party, Keith Cooper, FIFA's director of communications and someone the authors have done their best to like, makes it clear that he has been "deeply, deeply disappointed" with a piece the authors have written for the *Financial Times* on the first FIFA executive meeting under its new president. With this as a starting point, various written communications between the authors and Cooper about the offending article, along with a research proposal they submit to a British funding agency to study the governance and regulation of football, are used to illustrate how clashes of interests are often embedded in the power differentials that exist in the relationship between researchers and governing or funding bodies.

According to Sugden and Tomlinson (1999), governing bodies prefer noncontroversial, noncritical research, conducted and reported by "tame academics" in ways that promote a positive image to the outside world. In contrast, investigative and critical researchers seek a social scientific understanding that seeks to challenge the "official" view of events by looking behind the scenes at what really goes on in various domains. As Sugden and Tomlinson note, "it will not make you friends in high places; it will not ensure ask-backs" (p. 397). However, as their confessional tale reveals, for them these professional risks are worth taking.

Finally, there is an article by Kluge (2001) titled "Confessions of a Beginning Qualitative Researcher." Here, the author explores her reactions to the trauma of having to rewrite her doctoral dissertation not once, but twice. At the first meeting with her research consultant, Dr. Miller, Kluge goes in expecting a straightforward discussion of what she thinks is the final draft of her dissertation. Confidence that this draft will be the last soon evaporates when Dr. Miller recommends that Kluge rewrite *all* the 15 structural descriptions her dissertation contains.

"I'm *not* doing it again!" My pen slaps the tabletop; papers piled high quiver from the disturbance. I rise up out of my chair and bolt toward the window. The air feels charged with electricity—like it does in summertime in the high country before a storm. The last sunlight of the day strikes the tops of the buildings outside my room. I stare at the glow. As tears well up in my eyes, the buildings blur; my mind transforms them into mountains. Eleven stories up, I feel suspended, as though I am teetering on a narrow precipice several thousand feet above the ground. My body is trembling. . . . I see no handholds or footholds within reach. My voice—like those papers, still quivering, repeats, "I'm *not* doing it!"

Kluge, 2001, p. 329

With this as a starting point, Kluge's (2001) confessional chronicles her research odyssey, explaining how, as someone moving from quantitative inquiry into qualitative inquiry via phenomenology, she not only arrives at this crisis point but also works through the crisis and learns from it. Her journey is explicated using rock climbing as an extended metaphor not only because Kluge is an avid rock climber but also because she realizes that, for her, rock climbing and being a beginning qualitative researcher have similarities.

Kluge does the first rewrite. Her confessional tale (2001) deals with the struggles, uncertainties, and tensions involved in rewriting all 15 participant narratives, as well as the learning process involved, in an informed and reflexive manner. But worse is to come. Having produced a new draft, Kluge learns from Dr. Miller that she has overcompensated. Verbatim quotes dominate the structural descriptions; they are too similar and redundant to the textual descriptions. Dr. Miller recommends that Kluge undertake *another* rewrite.

What?! I had completely rewritten everything. I had diligently bracketed my preconceived notions of the experience of being physically active for a lifetime and "stayed true" to the participants' experiences by using their own words in the descriptions. I had written and rewritten the narrative three times. . . . I had written 15 participant portraits, 15 textural descriptions, and 15 structural descriptions—45 narratives in all! . . . Did I rewrite? Yes. Did I agonize over it? Absolutely.

Kluge, 2001, pp. 332-333

Thus, Kluge (2001) gets it "wrong" twice and has to rewrite her doctoral dissertation twice. On each occasion, her emotional reactions are laid bare for the reader to feel. These reactions are not dealt with in a self-pitying manner, however; rather, Kluge uses her feelings about events as a resource to better the world of phenomenological research and the process by which she has come to understand it better. Her confessional tale should provide invaluable guidance for both neophyte researchers and their dissertation supervisors as they embark on similar journeys.

Reflections

All too often, the political, personal, ethical, and messy realities of qualitative research are not formally documented. Rather, by the time the research is presented or written up, all the perils and pitfalls of the research experience have been omitted or smoothed out in a tidy report outlining what went right rather than what went wrong in a research endeavor. According to Boman and Jevne (2000), this is understandable given the desire to do well and achieve certain research goals.

> Depending on the circumstances, it is not always easy to write about the problems encountered due to something we or someone else may have done. However, by reporting only where things went right, the learning that comes from making mistakes remains private.
>
> *Boman & Jevne, 2000, p. 547*

Wolcott (1995) reminds us that reports of failed fieldwork are even more elusive than accounts of the political realities of fieldwork. This absence worries Wolcott; if the difficulties that lead to failure remain only in the private notes of the researcher, "then in the long run it may be of no more consequence than a forgotten summer holiday" (p. 226). As a consequence, the only reports available to the beginner will be those of successfully completed fieldwork projects. Wolcott notes that "the same few [accounts] are held up to generation after generation of students. As with other art worlds, we, too, place the masterpieces on so high a pedestal that they tend to inhibit when they are meant to inspire" (p. 226).

Wolcott (1995), like Boman and Jevne (2000), is right: Academic communities are not good at coming forward about their failures and mistakes. This truth is even more pronounced in emerging disciplines, like sport and physical activity, which aspire to the mantle of "scientific" credibility. As a consequence, realist tales, by definition and for appropriate reasons given their purposes, gloss over the ethical and methodological dilemmas that confessionals set out to expose. This exposing of fieldwork, of going behind the cleaned-up scenes, is therefore of vital importance if a reflexive and critical stance is to be adopted toward both the process and the products of the qualitative research enterprise. As such, confessional tales play a crucial role in highlighting the perils and pitfalls of the research experience, helping fellow inquirers learn from private mistakes and removing the inhibitions generated should novice researchers be fed solely on a diet of completed, methodologically sanitized, and successful research projects.

Given their pedagogical potential, M. Punch (1994) calls for the research community to tell more stories about actions and events that reveal the stress, the deep personal involvement, the role conflicts, the physical and mental effort, and the discomfort of qualitative research so that we might be better prepared for such realities in our own research endeavors. As such, the confessional tales considered in this chapter can act as a valuable resource for courses on qualitative research in sport and physical activity. This is particularly so given their ability to highlight fieldwork as a hermeneutic process and to raise a host of ethical and methodological questions about the basis of ethnographic authority—how we come to know about ourselves and others via our research activities. Indeed, these confessional tales go some way toward meeting the obligations outlined by M. Fine (1994) for qualitative researchers to ask messy questions about methods, ethics, and epistemologies (see chapter 1 of this book).

By skillfully working in the spirit of self-reflexivity, producers of confessional tales are able to provide a personal voice that can be a gift to readers. In these circumstances, Van Maanen (1988) notes, "the confessional becomes a self-reflective meditation on the nature of ethnographic understanding; the reader comes away with a deeper sense of the problems posed by the enterprise itself" (p. 92). Indeed, in Eisner's (1991) terms, the scholars focused on in this chapter have engaged in acts of "connoisseurship." That is, they have presented personal insights of the research act as part of an examination of their own research experience.

Of course, as Van Maanen is quick to point out, "In unskilled hands, a wild and woolly involuted tract is produced that seems to suck its author (and reader) into a black hole of introspection" (1988, p. 92). This dilemma for confessionals is recognized by Brackenridge, who notes, "There is also, of course, a danger of attempting to become 'more reflexive than thou' leading to one becoming trapped inside a circular reflexivity" (1999, p. 400).

There are other dangers for producers of confessionals. For example, Brackenridge is aware that her reflections might be viewed as narcissistic and "be seen as merely self-pity" (1999, p. 400). Likewise, Humberstone notes that by using her own experiences to raise issues she "could be charged with self-indulgence or even paranoia" (1997, p. 209). Certainly, there are some who find such tales distasteful in the extreme. For example, Sanders could not express his views of confessional tales more bluntly:

Post-modern ethnography-like exercises regularly also tend to be intensely narcissistic. As we know from our everyday experience, interesting people talking about themselves is quite interesting, while boring people talking about their thoughts and experiences can precipitate in the listener a desperate desire to escape. The self-focused "confessional tales" (Van Maanen, 1988) offered as the core of many post-modern ethnographies regularly go well beyond my own tolerance. . . . At worst, these works represent lengthy therapeutic rambling in which the writer insists upon telling us about his or her dreams, personal insecurities, "meditations," and sources of "panic." When confronted by work of this character one would reasonably conclude that the writer's time and energy would be better used if he or she expended more of them in therapy rather than thus assaulting innocent colleagues and taking up limited journal space. Certainly articles like this are NOT "sociology" unless one stretches this category to the breaking point.

Sanders, 1995, p. 96

Similar charges are often leveled against those who produce autoethnographies (see chapter 5) and, as with confessionals, are usually based on a misunderstanding of the nature of the genre and its purposes. Given that these charges are systematically refuted in the next chapter, I will not dwell on them here. As Van Maanen (1988) comments, "the reader who wonders why the confessional writers don't do their perverse, self-centered, anxiety work in private and simply come forward with an ethnographic fact or two are, quite frankly, missing the point" (p. 93).

At their best, confessional tales have much to teach us about the ethical and methodological complexities of the qualitative research process. If nothing else, they can serve to challenge the somewhat simplistic wisdoms espoused in many methods books, which tend to reduce qualitative research to a number of nonproblematic techniques. Finally, as Smyth and Shacklock emphasize, confessional tales have the potential to allow others to experience something of the struggle and excitement of the research act: "They give expression to inner dialogue that generally exists only as a sub-text in parts of the research account. . . . these reflexive accounts give valuable readings of researcher understanding of how complexity in research is understood by researchers themselves" (1998, p. 2). On this basis alone, the development of such tales deserves our attention in the future.

Chapter 5

Autoethnography

Autobiographical writing has a long and distinguished history and produces one of the most popular genres of books sold today. According to Stanley (1993), however, most analytical interest in this genre has been located outside of the social sciences and left to the domain of literary criticism. There are signs of change. For example, Morgan (1998), in his presidential address to the British Sociological Association, noted that, in recent years, considerable attention had been devoted to the uses of autobiography. Against this backdrop, some scholars in the social sciences have used their own experiences for analysis and have produced what have been called *narratives of self* or *autoethnographies*.

L. Richardson (1994) defines the narrative of self as an evocative form of writing that produces highly personalized and revealing texts in which authors tell stories about their own lived experiences. Dramatic recall, strong metaphors, vivid characters, unusual phrasings, and the holding back on interpretation invite the reader to emotionally relive the events with the author. In writing these frankly

subjective narratives, "ethnographers are somewhat relieved of the problems of speaking for the 'Other,' because they are the 'Other' in their texts" (p. 521).

Similarly, Ellis and Bochner (2000) define autoethnography as "an autobiographical genre of writing and research that displays multiple layers of consciousness, connecting the personal to the cultural. . . . Autoethnographers vary in their emphasis on the research process (graphy), on culture (ethnos), and on self (auto)" (pp. 739-740).

Commenting on what she called "heartful autoethnography," Ellis (1999) notes the following characteristics: the inclusion of researchers' vulnerable selves, emotions, bodies, and spirits; the production of evocative stories that create the effect of reality; the celebration of concrete experience and intimate detail; the examination of how human experience is endowed with meaning; a concern with moral, ethical, and political consequences; an encouragement of compassion and empathy; aid to help us know how to live and cope; the featuring of multiple voices and the repositioning of readers and "subjects" as coparticipants in dialogue; the seeking of a fusion between social science and literature; and the connecting of the practices of social science with the living of life.

For Ellis (1997, 1999), autoethnography as a way of knowing starts with her personal life and then pays attention to her physical feelings, thoughts, and emotions. Ellis uses what she calls *systematic sociological introspection* and *emotional recall* to try to understand the experiences she has lived through, and then she writes about these experiences using a variety of genres—short stories, poetry, fiction, novels, photographic essays, personal essays, journals, fragmented and layered writing, and social science prose (see Ellis, 1993, 1995a, 1995b, 1995c, 1998, 2001). The hope is that by exploring a particular life, the chances for understanding a way of life are enhanced.

The characteristics of and hopes for autoethnography are evident in the work of a growing number of scholars in the social sciences (for an extensive listing, see Ellis & Bochner, 2000). As an exemplar, B. Smith (1999) provides a narrative of self about his ongoing roller-coaster ride with severe clinical depression. According to Smith, realist tales of depression tend to represent the subjects' experiences from highly theoretical perspectives that close down alternative interpretations and bury people's voices beneath layers of analysis: "As a result, for me, they strip away the depth and intense emotional experience of the various depressions. People's words and worlds are flattened. Thus, a disembodied and emotionless account prevails" (p. 266).

In contrast, B. Smith (1999) uses three aspects of introspection, or self-ethnography (diaries and free writing, self-introspection, and interactive introspection), as resources to construct a messy text about his lived, embodied experiences of depression in a way that evokes the vulnerabilities, ambiguities, ongoing struggles, and gendered nature of this condition. By presenting his narrative of self in the form of short stories and poetry, Smith invites the reader to rupture the traditional pattern of scientific knowing and to "feel, hear, taste, smell, touch, and morally embrace the world of depressions" (p. 275).

Similar concerns over the effect of traditional forms of telling inform the autoethnography of Tillmann-Healy (1996) when she reflects on the complex relationship she has with her body, food, and bulimia—her secret life—in a culture of thinness. Her article begins as follows:

> In the spring of 1986, at the age of 15, I invited bulimia to come live with me. She never moved out. Sometimes I tuck her deep in my closet, behind forgotten dresses and old shoes. Then one day, I'll come across her—as if by accident—and experience genuine surprise that she remains with me. Other times, for a few days or perhaps a week or month, she'll emerge from that closet to sleep at my side, closer than a sister or a lover would. This is our story.
>
> *Tillmann-Healy, 1996, p. 76*

Our story is contrasted to *their* story of bulimia as told in the extensive research by physicians and therapists based on laboratory experiments, surveys, and patient interviews. Tillmann-Healy (1996) is not against the medical profession, and she acknowledges that their goal of curing this disease is an admirable one. She argues, however, that devoting all the attention to this end obscures the "emotional intensity of bulimic experiences and fails to understand what bulimia *means* to those who live with it every day and what it *says* about our culture" (p. 80). Tillmann-Healy points out that even though she is no "authority" on bulimia, she can show the reader via her story what no physician or therapist can because, "in the midst of an otherwise 'normal' life, I experience how a bulimic *lives* and *feels*" (p. 80).

Tillmann-Healy (1996) examines bulimia though systematic introspection, treating her own lived experience as the "primary data." In contrast to the medical professionals, who seek to move toward general conclusions, Tillmann-Healy moves *through* experiential particularity. Likewise, while physicians and therapists use terms such as causes, effects, and associations to try to explain, predict, and control bulimia, Tillmann-Healy uses evocative narratives to try to understand bulimia and to help others see and *sense* it more fully:

> They write from a dispassionate third-person stance that preserves their position as "experts." I write from an emotional first-person

stance that highlights my multiple interpretive positions. Physicians and therapists keep readers at a distance. I invite you to come close and experience this world yourself.

Tillmann-Healy, 1996, p. 80

Thus, there is a specific reason why Tillmann-Healy tells her story in a particular way.

By using short stories and poetry about episodes in her life, Tillmann-Healy (1996), like B. Smith (1999) in his narrative of self-involving depression, takes the reader on a harrowing, emotional, evocative, and insightful journey without resolution. Tillmann-Healy reveals the painful irony of living simultaneously in a culture of abundance and a culture of thinness. In so doing, her story implicates the family and cultural stories that encourage many young women and, increasingly, young men to relate pathologically to food and to their own bodies. In this context, the reader comes to understand that bulimia is not such an illogical "choice" as many might assume it to be.

In the autoethnographies provided by B. Smith (1999) and Tillmann-Healy (1996), the embodied selves and the multiple shifting identities of the authors in all their messy complexity are central to the stories being told. The manner of their telling allows us to know and understand in different ways the phenomena they focus on. Similar tales are beginning to emerge in sport and physical activity.

Autoethnographies in Sport and Physical Activity

With regard to the emergence of autoethnographies and narratives of self in sport and physical activity, there has been a strong Scandinavian influence. For example, a group of Scandinavian scholars have used what Laine (1993) and Sironen (1994) call "memory-work." Like the systematic sociological introspection and emotional recall described by Ellis (1997, 1999), this work entails the systematic exploration of personal experiences of sport and body histories. Such an approach recalls events from the author-as-researcher's life, and these events then form the core of a written narrative that facilitates interpretation from a variety of theoretical perspectives.

Accordingly, Kosonen (1993) focuses on memories of her own running body as a young woman growing up in Finland to explore issues of femininity, sexuality, the social norms of surveillance that constrain women's bodies, and how they might be challenged. For example, in comparing memories of her body in motion and free as opposed to being stationary, gazed upon, and constrained, she recalls:

When sprinting I had no time to think about my body—do I have a potbelly, are my breasts bouncing? When I was running fast, I felt light. I was not big and clumsy, but I was a fairy, then a part I could never have got in the school plays. . . . Afterwards it seems strange that I was distressed by the idea of having to walk in town with legs that felt like blocks of wood, but then was perfectly comfortable running in front of thousands of people. No wonder that I am still looking for that runner-girl who may be hidden somewhere inside me.

Kosonen, 1993, pp. 23-24

Likewise, Kaskisaari (1994) draws on her "rhythmbody" and her personal experiences of growing up as an athletic girl to focus on lesbianism as a female experience that allows some athletes to resolve the conflict between traditional female roles and their own sexual identity.

In contrast, Tiihonen (1994) explores the social construction of male identities in Finland during the 1960s and 1970s by drawing on memories of his own sport involvement and body experiences as they were shaped in relation to asthma. The article opens with a section called "The Cough" that focuses on an incident when Tiihonen was training at a camp for the national U16-football team in 1975:

My legs give way. Vision clouds. I head straight for the toilet. I hope no one noticed, especially the coach. Over the toilet bowl and hacking. Not just a hack but a rasping cough which throws up green sputum down the front of my sports shirt. Tearing lungs, but the cough won't stop. I lean against the toilet for support. I'm faint. . . . When the coughing finally eases off, my shirt front is covered in green slime and my chest hurts. I'm amazed. This cough has bothered me for weeks, but I didn't think it was so bad. I'm scared, though I don't dare show it. I don't tell the coach.

Tiihonen, 1994, p. 50

Tiihonen (1994) then incorporates other memories as he constructs a narrative of illness based on his intensely embodied experiences of asthma. In writing through these experiences, he produces a multilayered text that draws the reader into his world while simultaneously thematizing the body as anxious, instrumental, male, ambivalent, disciplined, and released via its inscription in hegemonic masculinity. Furthermore, as the emotional experiences that Tiihonen gained in sport unfold in the stories he tells, we begin to understand how sport has contributed to his own learning about social class, power, authority, social bonding, gender relations, and sexuality.

In a similar fashion, Silvennoinen (1993, 1994a, 1994b) draws on memories evoked by photographs of himself as a child to trace a personal history of his own body awareness, and he uses memories of his childhood heroes to explore the social construction of specific forms of masculinity in Finnish culture. He notes that the body's memory is surprisingly resilient and capable of self-recognition when given the opportunity as, for example, when looking through a photograph album. His stories also illustrate how many forgotten or gray areas in one's own body awareness from the past have a powerful impact on, and relevance to, currently held identities: "'Unfinished' childhood experiences continue to travel with us. And this may be felt even as freedom" (1993, p. 31).

Of course, this is not to suggest that the production of autoethnographies is the sole preserve of academics working in institutions of higher education. For example, one of the most powerful, interesting, and eminently readable insights into the world of male bodybuilding has been provided by Fussell in his 1991 book, titled *Muscle: Confessions of an Unlikely Bodybuilder*. The author explores the reasons he—an Oxford graduate in literature, working in the publishing industry, aged 26, 6 feet 4 inches tall, and weighing 170 pounds—entered into the world of hardcore bodybuilding. He vividly describes some of his experiences in this subculture over a four-year period as he put on 80 pounds of muscle. As Fussell makes clear at the start of his book, "The following is an account of my journey—what I did, what I saw, what I felt. . . . I sing of dreamers and addicts, rogues and visionaries. And I sing of my own solitary pilgrimage into this strange world" (p. 14).

Muscle is written in the form of a novel and makes no pretense to being "academic." There is no discussion of methodology, there are no edited chunks of interview data, and there are no references or overt theoretical perspectives to frame the main storyline or the reader's interpretation. Instead, Fussell (1991) relies on his dramatic and impressionistic recall of events and uses various literary conventions to create rich, colorful, and believable characters, such as Bamm Bamm, Vinnie, and Sweetpea. These characters are drawn together in a plot that revolves around Fussell's personal quest to protect and strengthen his fragile sense of self by building a muscular body. This transformed body is to act as armor to keep his fears of the outside world at bay and his anxieties about himself as a person locked inside.

As the reader is drawn into this tale via chapters with titles such as "The Genesis," "The Walk," "The Metamorphosis," "The Bunker," "The Move," "The Juice," "The Digs," "The Blitz," and "The Diet," dramatic insights are provided into the social meaning of muscle in the bodybuilding subculture, the part that muscle plays in the construction of a specific form of masculinity, and the ways in which this kind of masculinity is *performed* on a daily basis. Insights are also provided into the motivations that drive some men to give up their jobs, abandon their friends, train with weights for fours hours a day for six days a week, eat the equivalent of five six-course meals a day, and both ingest and inject themselves with massive doses of steroids. This obsession is seen as part of a quest for a hypermuscular ideal that can leave these men physically and emotionally devastated. In the concluding chapter, called "The Aftermath," Fussell reflects back on his journey of the self:

> But this shell that I created wasn't meant just to keep people at bay. After all, a can of Mace could do that. No, this carapace was laboriously constructed to keep things inside too. The physical palisades and escarpments of my own body served as a rocky boundary that permitted no passage, no hint of a deeper self—a self I couldn't bear. . . . But behind that huge frame and those muscular sets, I felt shut up in a kind of claustrophobic panic. Not flexing but drowning. I felt like an actor victimized by his own success, condemned to play a role again and again and again. A role I spent four years seeking out and perfecting, but a role I was no longer willing to play. . . . I became a bodybuilder as a means of becoming a caricature. The inflated cartoon I became relieved me from my responsibility of being human. But once I'd become that caricature, that inflated cartoon, I longed for something else. As painful and humiliating as it is to be human, being subhuman or superhuman is far worse.
>
> *Fussell, 1991, pp. 248-249*

Against this backdrop, a growing number of scholars in sport and physical activity have drawn on their personal experiences to explore issues relating to body-self relationships over time, identity construction, gender, sexuality, aging, impairment, disability, race, and ethnicity (see Bandy & Darden, 1999a; Denison, 1999; Duncan, 1998, 2000; Fernandez-Balboa, 1998; Innanen, 1999; Markula, 2002; Milchrist, 2001; E. Miller, 2000; Parrott, 2002; Rinehart, 1998b; Sandoz, 1997; Sandoz & Winans, 1999; Silvennoinen, 1999a, 1999b, 2002; Sparkes, 1996, 1999b, 2002; Sudwell, 1999; Swan, 1998, 1999; Tiihonen, 2002; Tinning, 1998; Tsang, 2000).

To give a flavor of the range and form of autoethnographic writing in sport and physical activity, I have selected two exemplars. I recognize that by only giving partial coverage to each, I am not able to do justice to either of them, but I hope that the taste provided will encourage the reader to follow up and read them in their entirety. The first autoethnography explicitly juxtaposes personal experience stories against the author's academic interpretations of those stories. The second example uses personal experience stories but allows them to stand by themselves so that the academic voice is silenced. Therefore, as we move between the two autoethnographies, there is a shift in the balance between telling and showing. *Telling* occurs when writers intervene in the narrative and suggest how they might feel about characters or interpret events. In contrast, *showing* involves the author's effacement, so that the characters act out the story and *reveal* things about themselves without the author proposing interpretations.

Exemplar 1: Let Me Tell You a Story: A Narrative Exploration of Identity in High-Performance Sport (Tosha Tsang, 2000)

Tsang was a member of the 1995 and 1996 Canadian National Rowing Teams and won a silver medal with her crew, the Women's Eight, at the 1996 Summer Olympic Games. In her article, Tsang allows both herself and her audience to explore a variety of nuanced ways of knowing oneself in and though elite sport—for example, as a woman, as an academic, and as a heterosexual Chinese-Anglo feminist.

Tsang's goals are to give the reader a taste of her experiences as an athlete and to (re)create herself in the telling of these stories. A major dilemma for Tsang is how to give the reader a sense of the progressive and varied relationship she has to the various identities she inhabits as part of their ongoing construction, which depends on time, context, and interaction with others. Also, how might she give the reader a sense of her lived experiences as ongoing, multiple, and changing?

In an attempt to resolve these dilemmas, Tsang tells multiple autobiographical stories. By telling them from a slightly different angle each time, she hopes to give a fuller picture of her experiences. Therefore, in telling her stories, Tsang calls on different voices. For example, there is the *experiential* voice through which different identities announce and foreground themselves while others regress. There is the *reflexive,* or *inner,* voice that critiques the other voices even as they are in the midst of their telling. There is also Tsang's formal, *academic* voice that provides the reader with her interpretations of her own experiential stories. Her strategy, then, is to start by telling a story about her experience and then follow this with her own sense-making story that flows from it. As a consequence, various voices intermingle in a multilayered text to draw readers in and out of Tsang's experiences, and in and out of her interpretations, by drawing on their own experiences.

The first experiential story Tsang tells is called "Hairy Legs." It is based on an incident in which one of her teammates on the Canadian National Rowing Team half-jokingly asks Tsang if she is going to shave her legs before the Olympics. This throw-away comment, along with an apparently lighthearted threat from two of her crewmates to tie her down and apply hair removal cream to her legs, sets in motion for Tsang feelings of panic, which are then followed by a multitude of different thoughts and emotions surrounding the subject of her legs.

> You see, I am a female. More precisely, I am Canadian female—and I am a Canadian female with hairy legs. I don't shave my legs and haven't for years, and I like it that way. As long as most of my teammates have known me, I have *never* shaved my legs. And really, questions or comments from rowing friends have rarely come up about my legs . . . until recently.
>
> *Tsang, 2000, p. 48*

Having told the story and allowed it to unfold in her experiential voice, Tsang then switches to her academic voice to inform the reader that it has multiple meanings. For example, Tsang suggests it is a story about gender, about femininity, and about pressures to conform to a conventional, North American standard of beauty. It is also a story of association. That is, as the rowers' appearance is taken to reflect on their country and other Canadians. Therefore, the women are expected to conduct themselves as representatives of their country and as female athletes—that is, to be *feminine* and shave their legs.

The story of "Hairy Legs" is also a story of abiding laws: of not standing out, especially as deviant. Conformity is the rule: "We train our bodies to it every time we get into the boat, striving for synchronicity of movement. We discipline our bodies to stay in line, both literally and figuratively. We are also disciplined to conform" (pp. 49-50). Furthermore, the story is about identities and identification, plus homophobia in sport. What Tsang does reflects back on her crewmates as part of a team: a single homogenized unit. In this sense, her hairy legs are threatening. Tsang imagines that some of her crewmates might be thinking that her hairy legs signal not only that she is lesbian, but that the whole crew might be lesbian as well. As such, "Hairy Legs" is about the negotiation of Tsang's identity in a situation where her identity is bound together with the other characters in her story.

Finally, Tsang argues, "Hairy Legs" is a story of the gaze of authority. Drawing on Foucault's (1977) notion of panopticism, where one is on display and is aware of it, Tsang notes that this awareness acts as a constraint. For her, as the Olympic Games approached, this gaze intensified and multiplied. Not only was the authoritative gaze of coaches and sport administrators focused on the crew members, but the eyes of the world were also now watching. According to Tsang, as the team received more and more media attention, the crew members developed a certain self-consciousness. That is, they reflected the gaze toward themselves, which increased its power even further: "This is what makes my hairy legs a worthy topic for discussion. At the same time, it is an individual concern, a group concern, a national concern, and a societal concern of identity" (p. 50).

Next, Tsang tells a contrasting story called "A Face of Difference." This story revolves around the reactions of rowers at a training camp when one of them, Cory, shows up at a workout wearing makeup.

> The difference is in her face, and it screams out at me. . . . It's all there: lipstick, eye shadow, eyeliner, maybe even some founda-tion? It seems so out of place, so obvious because it's so out of context—the context being a workout. I can't help but think of a clown's face as I rudely stare. Some of us can't resist and feel compelled to comment. I am one of these people who feels it "necessary" to at least ask her what the special occasion is.
>
> *Tsang, 2000, p. 51*

The rower wearing makeup explains that she has a full-time job to go to after the practice with little time to spare. Therefore, she applies water-proof makeup before she leaves home, goes to practice, takes a quick shower afterward, and goes directly to work—presentable and on time. This explanation satisfies Tsang, who sees Cory after practice in her formal business attire next to a car instead of a boat. In this context, the makeup fits in and Tsang feels bad about the thoughts she had earlier regarding Cory's makeup as garish and vain.

According to Tsang, "A Face of Difference," like "Hairy Legs," is a story about femininity. In this story, however, it is Tsang who is doing the pressuring. By asking Cory about her makeup, Tsang draws attention to something that is different and out of place. Tsang also passes judgment on the acceptability of Cory's reasons for wearing makeup to practice. Another difference in the story is that, even though it is about construct-ing femininity, it is not the conventional North American femininity. This kind of femininity is discouraged by Tsang and the other rowers in the particular context of a rowing workout since it may signify a lack of seri-ousness or an excess of vanity. Therefore, the ways in which femininity is *performed* carry meanings for those who are doing the performing and doing the watching. Thus, in nonworkout situations makeup is seen by many of the rowers as acceptable and is encouraged, but in a workout setting, different standards seem to operate regarding what is appropriate female conduct and appearance.

If we compare the two experiential stories told by Tsang and noted in her sense-making stories, it becomes apparent that not only are there differences in various femininities but also that power operates through and on these in different ways. In "Hairy Legs," Tsang is on the receiving end of the pressure to conform to a conventional form of femininity. On the other hand, in "A Face of Difference," Tsang is the one exerting the power. Her stories show how power dynamics change as contexts and meanings change. They also reveal that power roles, like identities, are not fixed but fluid.

> The telling of these two experiential stories allows us to see some of the contradictions and disruptions of the notion of identity as a unitary whole, as well as the power relation within which these identities are "crafted." . . . It also allows me to form sense-making stories, which, in turn, also influences what gets incorporated into my identity and what gets left out.
>
> *Tsang, 2000, p. 52*

Tsang then tells two more experiential stories, "Becoming Chinese" and "My Relations." The first centers on a time when she was singled out by a reporter for Canada's largest Chinese-Canadian newspaper as one of the few Chinese-Canadians in the Olympic team. The interview highlights a sense of difference for Tsang and she admits to feeling a "fraud." This is because Tsang knows nothing of Chinese-Canadian culture and does not speak Chinese or even know how to write her own name in Chinese. Indeed, Tsang is only half Chinese, and her father—who is "full" Chinese—was born and raised in Regina, Saskatchewan, and does not speak Chinese either:

> I figure they must have just seen my name on some athletes' list and concluded that since my last name was Chinese, I was part of their community. . . . As they [reporter and photographer] leave, I feel flattered that they identified me as "one of their own," yet I also feel I am an awkward white person trying to fit into a foreign culture.
>
> *Tsang, 2000, p. 53*

The next experiential story, "My Relations," is also about otherness. This story revolves around Tsang and her crewmates at an international rowing regatta, when they focus their gaze on the Chinese team. Members of the Canadian crew note that the Chinese team makes more noise than the Canadians do when they row. This reaction makes Tsang feel uncomfortable. Even though she finds the actions of the Chinese team as strange as her crew members find them, she still makes a connection between herself and the Chinese rowers. In part, this is because when she compares the Chinese crew to her own, the visual difference between them highlights Tsang's difference from her own crew:

> I look more like the Chinese in some ways than I do my crew mates, all of whom are (appear to be) white and many of whom are blond. I am white, sort of. I am also Chinese, sort of. I can pass for either, depending on who is looking. I am also the only one in my crew who makes "noise" when she rows. Despite myself, I feel a connection.
>
> *Tsang, 2000, p. 54*

This connection creates identity dilemmas for Tsang that are exacerbated when a crewmate jokes that the noisiness of the Chinese crew explains Tsang's own noisiness when she rows. This association troubles her in terms of the way the Chinese crew were "othered," and then Tsang was associated with them. This is because her crewmates also "othered" Tsang and made her feel "not wholly part of 'us' in that moment, a part of me is foreign, strange (the part of me that grunts loudly—something that no one else on my team does; something different)" (p. 54).

In the sense-making story that reflects on "Becoming Chinese" and "My Relations," Tsang notes that they are stories of racialization and identities. They are about the multiple identities she inhabits or has ascribed to her by others and the interplay, plus tensions, between them. They also openly challenge and problematize the normalized racelessness (i.e., white) that frames so many stories of athletics, both for Tsang and for the reader: "Both the untold story of white and the story of Other (within and without me) find expression through the multiplicity of my identities. Yet how and when and which will be expressed is not an easily predictable matter" (p. 55).

Tsang continues to foreground her academic voice in the next two sections, "There Is Method to This Madness" and "The End . . . ?" The former discusses the nature of narrative tellings and how readers come to know through them. Here, Tsang emphasizes that the importance of the stories she (re)tells are in what they signify and conjure up for the reader. As she explains, "I have no absolute answers; I can only share with you my experiences and hope that they will have some meaning for you, represent something to you, but not necessarily exactly what I experienced" (p. 55).

The latter section discusses the multiple voices that Tsang has drawn on in her stories and the way in which this has allowed her to ask questions about identities (racialized, gendered, classed, and embodied) and multiplicity. For Tsang, these issues have been addressed not only in content but also in form, "which has provided me with a new way of seeing and, thus, knowing identities" (p. 57). But, of course, this knowing is not and never will be complete or conclusive. Therefore, to be consistent with the idea of identity as process and the idea that content and form are connected so as to reflect each other, Tsang ends and begins with another personal experience story called "Growing Small." This story tells of how, as she grew up, she felt herself to be a tall athletic girl, but when she joins the Canadian National Rowing Team, despite being a tall athletic woman, she starts to be labeled "small." The story explores Tsang's reactions to this labeling and closes when she leaves the rowing scene and starts "to grow taller" (p. 57).

Exemplar 2: Reflex: Body As Memory (Margaret Carlisle Duncan, 2000)

This article shows rather than tells via a series of nonfictional vignettes based on Duncan's autobiographical recollections of her body, for herself and in relation to others, over time and in different contexts. Throughout, Duncan's academic voice is stilled. No citations, no references, and no scholarly interpretations are provided. As such, the article is even more open to multiple interpretations relating to, for example, the role of sport and physical activity in the ongoing construction of gendered identities during the life course.

The first vignette, "Learning to Swim," takes place when Duncan is four years of age. She learns first by herself in an unstructured situation that allows her to experience the sensuality of the sea.

> The water is warm as blood, and it pulses too. . . . I've discovered something magical about myself: I can float! Each time a wave crests, I allow myself to be buoyed upward and pump my arms in the water. I am vertical, a gyroscope spiraling slowly through space. After a bit, I remember that when people swim, they're horizontal. I cautiously tilt forward, imagine my toes growing webs, flutter my arms, then my legs. . . . I propel myself forward a few feet—cowabung! I try it again, salty water stings my eyes and slops into my mouth, but I don't care, I'm on a roll: I taste pure freedom, pure happiness.
>
> *Duncan, 2000, p. 60*

This experience is juxtaposed with the formal swimming lessons that Duncan's mother sends her to a few years later. Here, correct stroke technique is taught by a teacher Duncan does not like. Duncan decides that the breaststroke is for sissies; she favors the crawl, in which she can swim laps for hours "stroking toward oblivion. (This enchanted state is what drives me, as an adult, to the swimming pool)" (p. 60). Duncan is praised by her teacher and her mother on the progress she makes, but she herself has no sense of practicing skills: "I just pick them up somehow when I'm in the water" (p. 60).

Following one swimming lesson, one of Duncan's friends shows her a place in the basement window where they can see into the boys' locker room: "All those tiny, flopping peenies are captivating in a gross kind of way. We see some grownups approaching and scatter, giggling like maniacs" (p. 61).

The next vignette is called "Horseback Riding." Duncan is now in the fourth grade and is on a trail ride with some school friends. An inexperienced riders, she sits astride a large horse called Samson and is "fearful and proud in equal measures" (p. 61). Duncan is intimidated by Samson's strength and power. Her lack of skill as a rider, along with her need to save face in front of the group, eventually leads to a fall that renders her unconscious and suffering from concussion when she comes to. Once her memory of the trail ride returns, Duncan has a perverse compulsion to ride again.

> I am relentless. I pester my parents until they finally give in, and each Saturday my mother drives me out to Poolesville for English riding lessons. It takes 45 minutes to get there; I am terrified the whole time up to and including the actual lesson, can feel the skin on the backs of my hands prickle with tension. Afterwards, though, I feel wonderful, really accomplished, and forget my terror until the next Saturday
>
> *Duncan, 2000, p. 61*

"Acquiring Muscles" is the next story. Duncan is now 11, in her "tomboy phase," and resents wearing the mandatory dress for school. Resistance begins by wearing shorts under the dress so that she can perform acrobatic tricks during recess without letting the boys see her underpants. For her birthday, Duncan goes with her parents and buys a baseball bat, a ball, and a fielder's glove. Her father gives her technical advice on what to buy and how to season the glove. Duncan and her friend Steff inaugurate new glove games, and they play baseball with whomever they can find: "I enjoy bringing my own sporting equipment—my tomboy's identification kit" (p. 62).

Duncan also arm wrestles and performs gymnastic tasks with Steff. They get stronger. Her muscles begin to assert themselves: "In my bedroom, I clench my fist in front of the mirror and admire the peak of my muscle. I'm wearing a sleeveless shirt to give my arm maximum exposure" (p. 62). Proud of her emerging muscularity, Duncan decides to show her mother and brother. She is floored and feels betrayed when her mother suggests that muscle on women is unfeminine, and her brother insinuates she is trying to look like a boy. Duncan is hurt and angry: "I think I will sock my brother in the stomach, try to knock the wind out of him, but instead I burst into tears, run into the house, and fling myself on the bed" (p. 62).

In the story called "Playing Basketball," Duncan and her friend Annie, who is 14 years old, perfect their lay-up shots plus several variations in her parents' driveway. They also play "Horse" but are not competitive in this game, allowing each other to win and lose in equal measure: "Out of good manners—we are southern girls, after all" (p 63). Their basketball skills honed, they continue their after-dinner games on going to junior high. Then, in a physical education lesson, Duncan gets to play a game of basketball:

> I don't have much stamina, but I can shoot from anywhere on the court and score points. It's thrilling to be so good. . . . The basketball is an extension of my will, my arm. The way it swishes though the net gives me almost visceral satisfaction.
>
> *Duncan, 2000, p. 630*

Unfortunately, there is no girls' basketball team at her school, and Duncan has to make do watching the boys compete outside of gym class, even though she knows they would win games with her playing. In contrast, after three weeks of archery classes, Duncan never once hits the target: "But it doesn't even matter to me because I know, *I know* I am queen of the basketball court" (p. 63).

"Running" is the next story. Duncan is now married and follows her husband into his first full-time academic post at a time when she realizes she is not going to make a career as a professional musician. Duncan takes a job as secretary to the head of a kinesiology department at Purdue University. Here, Duncan is surrounded by people who exercise at lunchtime: "My only exercise since high school has been swimming, and then only occasionally. . . . Running is *not* my idea of a good time. I think of it as punitive" (p. 64). Despite her fears and worries about running, a friend encourages her to start, and, over time, she gradually gets to enjoy the experience: "Running makes me feel purged and holy. I like listening to my rhythmic breathing, feeling my legs piston around the track, every body part moving in perfect synchrony. I even like the sweat and the pleasant sense of fatigue afterwards" (p. 64). Pregnant with her first child, Duncan continues to run though her fourth month, despite the pain running causes, before she turns to swimming. She counts the days until she can begin running again.

In "Becoming a Martial Artist," Duncan has moved to Milwaukee, where she and her husband teach at the University of Wisconsin. She now has two daughters aged 10 and 8, and the three of them attend an evening family class in Tae Kwon Do. During the class, the teacher, Master Cho, asks students to tell him the meaning of a green-belt form as part of their preparation for a grading. Those he asks cannot remember, but Duncan can and Master Cho nods approvingly. At the grading for her red belt, Duncan has to break some boards using a variety of techniques. The boards get progressively larger as students proceed though the belt hierarchy, but at every level, women are given smaller boards to break than men. For her grading, Duncan requests the boards that the men break. Then, using first a flying side-kick and then a reverse arm strike, she breaks the boards cleanly: "I beam at the panel of judges. At the end of testing, the judges confer. Master Cho stands up and announces in authoritative tones, 'Red-belt women did *good job*. They have *fighting spirit*.' He glances in my direction" (p. 66).

The final story is called "Aerobicizing." After a spate of torn muscles, Duncan concludes that Tae Kwon Do is no longer the best exercise for her and that she needs to try something different. Aerobics is a possibility, except that Duncan, having checked out a Jane Fonda video, sees this as a "kind of sissy thing. *Not* for the lion-hearted of this world. Designed for people who don't like to sweat. . . . Aerobics? I don't *think* so. Too many smiling meat puppets" (p. 66). Encouraged by a colleague who attends aerobic classes three times a week, however, Duncan decides to go along. Once she's there and involved, a host of her preconceptions and prejudices about the nature of aerobicizing and the people who attend such classes are challenged: "Not [a] single bimbette in the bunch. I permit myself to stop glowering and start feeling more comfortable about being part of this group" (p. 67). Duncan also comes to realize that the exercises are strenuous and require coordination. At the end of the class, she admits to the teacher, "It's nothing *at all* like I imagined" (p. 68). The next morning, her body feels the effects of her exertions, but she hints that she will continue to go to classes.

On the Charge of Self-Indulgence

I hope that the extracts in the previous section give some flavor of how autoethnographies operate. Of course, this flavor is not to everyone's liking, and many in the academy adopt a suspicious and hostile stance toward the autoethnographic venture. According to Krizek (1998), "Many of us 'do' ethnography but 'write' in the conservative voice of science. . . . we often render our research reports devoid of human emotion and self-reflection. As ethnographers we experience life but we write science" (p. 93). In a similar fashion, Ellis and Bochner (2000) note that most writing in the social sciences is in the third person, passive voice—as if written from nowhere by nobody. They argue that traditional conventions militate against personal and passionate writing: "Once the anonymous essay became the norm, then the personal, autobiographical story, became a delinquent form of expression" (p. 734).

By writing themselves into their own work as major characters and choosing to foreground their own voices, the scholars mentioned in this chapter have challenged accepted views about silent authorship and author-evacuated texts. Most often, as Charmaz and Mitchell (1997) argue, scholarly writers are expected to work silently on the sidelines and to keep their voices out of the reports they produce. In many ways, they are expected to emulate Victorian children. That is, to be seen (in the credits) but not heard (in the text). As such, "silent authorship comes to mark mature scholarship. The proper voice is no voice at all" (p. 194). Therefore, those who write autoethnographies and narratives of self in the domain of sport and physical activity have begun to produce what Church (1995) calls *forbidden narratives*.

As a form of forbidden narrative, autoethnography provides challenges to conventional ways of writing and knowing about the social world. Because it blurs boundaries, it has been subjected to criticisms. Probably the most common charge is that autoethnographies are simply exercises in self-indulgence (see also chapter 4 on confessional tales). For example, Coffey (1999) notes, "some would say that such texts are not 'doing' ethnography at all, but are self-indulgent writings published under the guise of social research and ethnography" (p. 155). Elsewhere she asks, "are we in danger of *gross self-indulgence* if we practice autobiographical ethnography?" (p. 132; my italics). This is a serious charge and, even though various criteria for judging different kinds of tales are covered in detail in chapter 9, it seems appropriate that the universal charge of self-indulgence be dealt with, and refuted, at this juncture.

Mykhalovskiy asks,

What is it about personal narratives in social science that people find so offensive? It must be something more than a challenge to scientific objectivity etc. There is something I think that people really despise about this kind of work that I haven't quite figured out.

Mykhalovskiy, personal communication, April 2000

Certainly, those that produce autoethnographies are acutely aware of this hostile atmosphere. For example, in an autoethnography that explores his thoughts about the powerful imprint left by a father who committed suicide when his son was only 10 years old, Gray expresses the following concern:

> Perhaps my biggest struggle throughout the writing has revolved around taking what has always been very private and making it public. Despite my determination to make this happen, I hear voices that tell me this is a bad, dangerous course to take. These voices say, "This work is narcissistic, and self-indulgent, and you are embarrassing yourself though a melodramatic, emotional self-exposure." When I read the social science literature on masculinity, my fears are confirmed. I encounter writers who express a disdain for autobiographical approaches, an implied suggestion that the genre is populated with narcissistic egomaniacs. Is that me?
>
> *Gray, 2000b, p. 110*

Of course, producers of autoethnographies and narratives of self (like any other pieces of research and forms of representation) need to be aware that their writing *can* become self-indulgent and masturbatory rather than self-knowing, self-respectful, self-sacrificing, or self-luminous. As Pelias (1999) recognizes, producers of poetic essays do risk the appearance of self-indulgence. They can seem unbridled as they attempt to pull personal experience into the scholarly equation. Furthermore, as Pelias acknowledges, drawing on the thoughts of Trinh Minh-ha (1989), there will be times when self-consciousness may lead to self-absorption and the author fails to land on the narrow and slippery ground between the twin chasms of navel-gazing and navel-erasing. Certainly, landing on this middle ground is no easy task.

Furthermore, as Hertz comments, "revealing oneself is not easy. For example, how much of ourselves do we want to commit to print? How do we set the boundary between providing the audience with sufficient information about the self without being accused of self-indulgence" (1997, p. xvi). Indeed, as part of his critique of the new fashion of self-revelation in what he calls narrative nonfiction in literature, Morrison comments, "Confessionalism has to know when to hold back. . . . It takes art. Without art, confessionalism is masturbation. Only with art does it become empathy" (1998, p. 11).

Certainly, the cautions provided by Hertz (1997), Morrison (1998), and Pelias (1999) need to be reflected on by all those who produce autoethnographies. The careful position adopted by these scholars, however, is a long way from the position so often adopted in the academic community, which tends to define all such productions as universally self-indulgent, regardless of the qualities of any individual venture. The implementation of this charge in a universal manner is, for me, a dangerous and threatening move that needs to be challenged and rejected.

According to Rinehart, the universal charge of self-indulgence leveled against autoethnography and other forms of vulnerable writing is based on a misapprehension of these genres as self-conscious navel-gazing and is "grounded in a deep mistrust of the worth of the self" (1998a, p. 212). This is because, as Krieger argues, the traditional view of social science is premised on "minimizing the self, viewing it as a contaminant, transcending it, denying it, protecting its vulnerability" (1991, p. 47). This contaminant view, according to M. Fine et al., demands that the researcher's self is something to be separated out, neutralized, minimized, standardized, and controlled: "This bracketing of the researcher's world is evident in social science's historically dominant literary style, which is predicated on a 'clarion renunciation' of the subjective or personal aspects of experience" (2000, pp. 108-109). Here, researchers are expected to ask participants to reveal their vulnerabilities but reveal nothing or little of themselves.

For Mykhalovskiy (1996), once the view of the social scientist's self as a contaminant is in place, then the way is opened to define any writing about one's self as self-indulgent. As a form of academic gatekeeping, he notes, "naming the work of a writer as work that indulges only that writer's self is peculiarly silencing. . . . one who is writing, above all else, writes about the other" (pp. 135-136). As Bochner and Ellis (1996b) point out, narcissism and related criticisms, such as self-indulgence and self-absorption, function to reinscribe ethnographic orthodoxy and resist change. Unfortunately, this regulation contains misplaced assumptions that need to be highlighted and challenged.

Challenging Misplaced Assumptions

In relation to autobiographical texts, Freeman notes that such work "might be of value to someone besides ourselves" (1993, p. 229). Likewise Church (1995), in exploring the personal, private, and emotional dimensions of research, notes how this work challenges male-dominant conventions concerning what can be discussed in academic settings. She emphasizes that including the emotional does not wipe out the public, the theoretical, and the rational. Rather, Church suggests, what we experience and present of ourselves as subjective or personal is simultaneously objective and public:

> I choose to foreground my own voice. This is not narcissism; it is not an egocentric indulgence. . . . Critical autobiography is vital intellectual work. . . . The social analysis accomplished by this form is based

on two assumptions: first, that it is possible to learn about the general from the particular; second, that the self is a social phenomenon. I assume that my subjectivity is filled with the voices of other people. Writing about myself is a way of writing about these others and about the worlds which we create/inhabit. . . . Because my subjective experience is part of the world, the story which emerges is not completely private and idiosyncratic.

Church, 1995, p. 5

Mykhalovskiy (1996) also questions the goals of autobiographical sociology. He argues that the claim of narcissism rests on an individual/social dualism that obfuscates "how writing the self involves, at the same time, writing about the 'other' and how the work on the 'other' is also about the self of the writer" (p. 133). Consequently, to characterize autobiographical work as self-indulgent is to make claims about its content by invoking a reductive practice that asserts the autobiographical to be only about the self of the writer and no one or nothing else. This kind of dualistic thinking, according to Jackson, is wasteful because it "wrenches apart the interlocking between self and society" (1990, p. 11).

In a similar fashion, Stanley (1993) notes that people do not accumulate their life histories in a social vacuum. That is, even though individuals may largely control the processes of recalling and interpreting past events, this process is also a social activity influenced by the people with whom the individuals interact. Therefore, the autobiographical project disputes the normally held divisions of self/other, inner/outer, public/private, individual/society, and immediacy/memory. Likewise, Gergen's (1999) social constructionist view of the self as *relational* challenges the dominant ideology of the self-contained individual that underpins notions of self-indulgence.

Drawing on the work of Mikhail Bakhtin, Gergen (1999) argues for a vision of human action in which rationality and relationship cannot be disengaged. Here, our every action manifests our immersion in past relationships and, simultaneously, the stamp of the relationship into which we move. Thus, for Gergen, any performance, such as writing an autoethnography, is relationally embedded:

Recall the way in which our expressions gain their intelligibility from a cultural history. In the same way that I cannot make sense if I use a word that I myself have made up, my actions will not make sense if they do not borrow from a cultural background. Thus, when I perform I am carrying a history of relationships, manifesting them, expressing them. They inhabit my every motion. . . . we are always

92

addressing someone—either explicitly or implicitly—within some kind of relationship.

Gergen, 1999, p. 133

Eakin (1999) also talks of relational selves and relational lives as part of his challenge to the myth of autonomy in autobiographical work. He warns against thinking of autobiography as a literature of the first person, since the subject of autobiography to which the pronoun "I" refers is neither singular nor first. In demystifying the notion of autonomy, he asks the following:

Why do we so easily forget that the first person of autobiography is truly plural in its origins and subsequent formation? Because autobiography promotes an illusion of self-determination: *I* write my story; *I* say who I am; *I* create my self. The myth of autonomy dies hard, and autobiography criticism has not yet fully addressed the extent to which the self is defined by—and lives in terms of—its relations with others . . . *all* identity is relational.

Eakin, 1999, p. 43

Against this backdrop, Mykhalovskiy challenges reductive practices and argues that "to write individual experience is, at the same time, to write social experience" (1996, p. 141). Indeed, he argues that making connections between individual experience and social processes, in ways that point to the fallacy of self/other, individual/social dichotomies, is the task to which autobiographical sociology is best suited. As one of the reviewers noted about my autoethnography published in 1996, "The Fatal Flaw," "Especially, noteworthy are his reflections on a number of 'blurrings' (disease-illness, mind-body, public-private, man-woman, body-emotion, physical prowess-intellectual strength, self-others, conformity-deviance, etc.)" (Sparkes, 2000, p. 25). Likewise, in reflecting on her own narrative of self, Tsang notes:

I have claimed these stories to be my own, yet a story of myself, of my identity, necessarily involves and depends upon a story of the Other too. So these stories belong to them as well (albeit not in the same way or invoked with the same power)—the Other being the characters in the stories with whom I interact and compare myself and allude to. These are also the readers' stories, for through reading, readers construct their own meanings and identify with or resist certain elements of the story. How they do so not only reflects back on them and their own values and notions of themselves, but also implicates them as collaborators in the creation of the meaning of the text.

Tsang, 2000, p. 47

These views receive strong support from Bochner and Ellis, who ask, "If culture circulates through all of us, how can autoethnography be free of connection to a world beyond the self?" (1996b, p. 24). They rightly conclude that it cannot, and so the concerns that some critics have raised about the self-indulgence of autoethnographers become absurd.

Another question Mykhalovskiy (1996) asks is: "To whom does the autobiographical text speak?" Condemning autobiographical texts as self-indulgent, he suggests, reflects the experience of a particular reader that is then generalized to a universal reader. "Here, the charge of self-indulgence is a contradictory reading, which as a specific or particular response invokes a universal reader who shuts out the possibility of the text speaking to others" (p. 137). Such a move also seems to imply that the universal reader has universal characteristics and always reads from the same social positioning. These assumptions may inform the production of some texts, but they do not hold for autoethnographies that call on different ways of telling and showing and that invite a different kind of reading.

From Passive to Active Readership

Barone (1990) notes how paradigmatic texts are designed in accordance with a logico-scientific mode of thought; invoke a universal, passive, unengaged reader; and call for *efferent* readings that focus on the concepts, ideas, and facts to be retained and the actions to be performed as a result. Such texts have the characteristics of an *industrial tool*, which is not meant to be dismantled and reconstructed as its function is to create a seamless, denotative, linear discourse that rearranges the relationship among complex phenomena into a propositional form. According to Barone, the text-as-tool does not prize metaphorical aptness but offers the standard of technical precision: "It is not designed to surprise the reader-as-user. Its modes of fashioning are not designed to challenge the common order. . . . This text offers one verbal version of reality, meant to be taken literally, taken for the only world that can be represented, the real one" (pp. 315-317).

In contrast, as a way of knowing, narrative implies a *relational world*. As McLeod (1997) notes, a story always exists in a space between teller and audience. The story may be created by the teller, but it is always created in relation to a particular audience, "so it is as if to some extent the recipient(s) of the story draw it out of the teller. . . . Even a story written alone, such as a novel, has an implied audience" (p. 38). Accordingly, Barone (1990) makes a strong case for literary narrative texts, which call on narrative modes of knowing, as an occasion for conspiracy that encourages the reader to engage in the activities of textual re-creation and dismantling. Here, reading is not a passive but an active process that people undertake from multiple positions. This point is emphasized by Tsang (2000):

> Even with my stories in print and susceptible to readings and rereadings of the same words, form, and medium, each reader brings different resources to a text and, thus, different tools for making meaning out of my stories. For example, a reader who identifies racially as being white may have a different reaction and set of experiences with which to refer to when reading my stories of racialization than does a reader who strongly identifies as a racial minority. This in turn may be different from someone who identifies as racially mixed, or someone who hasn't really thought about race at all. The stories may summon different experiences from different readers in a variety of ways and with each reading.
>
> *Tsang, 2000, p. 55*

Therefore, according to Barone (1990), an *aesthetic* reading of the text is called for in which the readers' attention is centered directly on what they are living though during their relationship with that particular text. Similarly, Frank (1995) suggests that readers of personal accounts of illness might like to think and feel *with* the story being told rather than *about* it. In distinguishing between the two, Frank comments,

> To think about a story is to reduce it to content and then analyze that content. Thinking with stories takes the story as already complete; there is no going beyond it. To think with a story is to experience it affecting one's own life and to find in that effect a certain truth of one's life.
>
> *Frank, 1995, p. 23*

Furthermore, as Bochner points out:

> We can call on stories to make theoretical abstractions, or we can hear stories as a call to be vigilant to the cross-currents of life's contingencies. When we stay with a story, refusing the impulse to abstract, reacting from the sources of our own experience and feelings, we respect the story and the human life it represents, and we enter into personal contact with questions of virtue, of what it means to live well and do the right thing.
>
> *Bochner, 2001, p. 132*

Therefore, as Bochner and Ellis note, a good account is able to inspire a different way of reading: "It isn't meant to be consumed as 'knowledge' or received passively. . . . On the whole, autoethnographers don't want you to sit back as spectators; they want readers to feel, care and desire" (1996b, p. 24). As such, authors of autoethnographies seek to produce writerly rather than readerly texts. The latter leads the reader logically, predictably, and usually in a linear fashion through the research process. Little space is available for readers to make their own textual connections between the stories and the images presented. In contrast, the writerly text is less predictable. It calls on readers to engage with the text and to bring to the reading their experiences.

This kind of thinking fits with poststructuralist views that stress the interaction of the reader and the text as a coproduction, and reading as a performance (see chapter 1). As part of this performance, readers must be prepared to make meaning as they read, put something of their own into the account, and do something with it. To this end, according to Barone (1995), the artfully persuasive storyteller—who trusts the reader and understands the necessity of relinquishing control, of allowing readers the freedom to interpret and evaluate the text from their unique vantage points—will coax the reader into participating in the imaginative construction of literary reality through carefully positioned *blanks* in the writing. These invite the active reader to fill them with personal meaning gathered from outside the text. Here, the aim of the storyteller "is not to prompt a single, closed, convergent reading but to persuade readers to contribute answers to the dilemmas they pose" (p. 66).

In reflecting on how she came to write *Final Negotiations* (Ellis, 1995a), Ellis makes the following points:

> My open text consciously permitted readers to move back and forth between being in my story and being in theirs, where they could fill in or compare their experiences and provide their own sensitivities about what was going on. I attempted to write in a way that allows readers to feel the specificity of my situation, yet sense the unity of human experience as well, in which they can connect to what happened to me, remember what happened to them, or anticipate what might happen in the future. I wanted readers to feel that in describing my experience I had penetrated their heads and hearts. I hoped they would grapple with the ways they were different and similar to me.
>
> *Ellis, 1997, p. 131*

This focus on reader response encourages connection, empathy, and solidarity, as well as emancipatory moments in which powerful insights into the lived experiences of others are generated (Sparkes, 1994a, 1997b).

Accordingly, a valuable use of autoethnography is to allow another person's world of experience to inspire critical reflection on one's own (Bochner and Ellis, 1996b; Ellis, 1997). Here, readers recontextualize what they knew already in light of their encounter with someone else's life. This may not always be a pleasant experience. When an autoethnography strikes a chord in readers, it may change them, and the direction of change cannot be predicted. Indeed, as one anonymous reviewer commented on my own autoethnography, "The Fatal Flaw" (Sparkes, 1996):

> In "tapping" and evoking different levels of experience / subjectivity he constructs a multi-layered text which allows for, rather than speci- fies, a wealth of insights reaching well beyond the author's particu- lar predicament. It makes you think and feel, and opens up a wide range of questions able-bodied people probably never think about. Actually, this text could be used as a great "sensitizing" agent in the classroom.
>
> *Sparkes, 2000, p. 25*

Besides having a pedagogical function, well-crafted autoethnographies can also act as a sacrament and a call to witness.

Sacrament and Witness

According to L. Richardson (2001a), autoethnographies can be a *sacrament*. To her this means two things:

> Experiencing the flow of writing *and experiencing connectedness to oth- ers.* The sense of time and space as separate is undermined, re- understood as deeply interrelated. As you write, you can find your- self connected to others; the meaning you construct about your life connects you to others, making communion—community—possible.
>
> *L. Richardson, 2001a, p. 37*

Autoethnographies can also act as a call to *witness* for both the author and the reader. For example, in chronic illness, as Frank noted, becoming a witness means assuming "a responsibility for telling what happened. The witness offers testimony to a truth that is generally unrecognized or suppressed. People who tell stories of illness are witnesses, turning ill- ness into moral responsibility" (1995, p. 137). Frank distinguishes between the witness in the traffic court and the illness witness. Both speak on the authority of being present, but the latter's testimony is less of seeing and more of being; "illness stories are not only about the body but *of* and through the body" (p. 140). This testimony, according to Frank, implicates others in what they witness. Witnessing is never a solitary act, and it always implies a relationship. Ill persons tell themselves stories all the time, but they cannot testify to themselves alone; "part of what turns stories into testimony is the call made upon another person to receive that testimony" (p. 141).

For Ropers-Huilman, acts of witnessing occur "when we participate in knowing and learning about others, engage with constructions of truth, and communicate what we have experienced to others" (1999, p. 23). These acts can have powerful consequences:

> Witnessing affects one's persona in its entirety—our bodies, hearts, and souls are changed and renewed by what we witness in our lives. . . . Witnessing is powerful. There are great opportunities and dangers inherent in the process of witnessing others' lives and constructing meanings about those experiences.
>
> *Ropers-Huilman, 1999, p. 24*

Witnessing also has *obligations* (Ropers-Huilman, 1999). These obligations include recognizing our engagement in active, yet partial, meaning-making; recognizing that, as witnesses, we will change others and our roles as change agents need to be considered with great intentionality and sincerity; being open to change; telling others about our experiences and perspectives; listening to the interpretations of other witnesses; and finally, exploring multiple meanings of equity and care and acting to promote our understanding of these concepts.

Of course, writers of autoethnographies are also called to act as witnesses to their own stories once they have been produced. According to L. Richardson (2001a), we do not live lives; we live plot-lines. Thus, writing autoethnographically can lead us to discover new things about our selves and our world: "We have the possibility of writing new plots; with new plots come new lives. One's life always exceeds the cultural script for it; we are not cultural clones" (p. 37). Thus, for the author comes the possibility of first recognizing entrenched cultural narratives and dominant "master" narratives, and then rejecting them by writing in ways that both resist and challenge the accepted norm. As part of this process the author might ask, What are the stories that currently shape my life? How do they constrain and empower me? How do they constrain and empower others? Do I want to change? What might be the new story? What might be the plot-line? How can I resist and alter the old ones? How do I change? How can my stories connect to those of others to become part of a new collective story? (see L. Richardson, 1990).

For example, as an active reader of my own autoethnographies, I have been forced to ask many difficult questions (see Sparkes, 2002). In what ways have I colluded with various forms of hegemonic masculinity, and how has this adversely shaped not only my own life experiences but also the experiences of those I am connected to? I am forced to ask, following Frank (1995), questions about *body-relatedness*. Do I *have* a body, or *am* I a body? I also have to ask questions about *other-relatedness*. For example, just what is my relationship, as a body, to other persons who are also bodies? How does our shared corporeality affect who we are, not only *to* each other, but more specifically *for* each other?

In grappling with these questions, writing autoethnographically has helped me to recognize better the stories, or plot-lines, interwoven over time into my own life and the lives of other men, which have shaped the construction of a specific kind of male who has been, and still is, encouraged to make significant emotional investments in a certain kind of body at the expense of other kinds of bodies. Thus, the call to witness can operate as a stimulus for change at the individual level. Of course, this does not guarantee change, but it is a starting point. As Frank notes, the moral imperative of narrative ethics "is perpetual self-reflection on the sort of person that one's story is shaping one into, entailing the requirement to change that self-story if the wrong self is being shaped" (1995, p. 158).

In light of the issues I have raised, I believe that the universal charge of self-indulgence so often leveled against autoethnography is based largely on a misunderstanding of the genre in terms of what it is, what it does, and how it works in a multiplicity of contexts. At its best it is able to provide access to the multiple subjectivities of social life and a range of embodied feelings, emotions, and reactions to others. Accordingly, this genre is able to raise sociopsychological questions that connect individual and group interaction to the surrounding social structures. Furthermore, as I have illustrated, autoethnographies can encourage acts of witnessing, empathy, and connection that extend beyond the self of the author and thereby contribute to sociological understanding.

The lack of understanding surrounding autoethnography and how it might be judged is not surprising. As DeVault (1997) notes, sociologists are not accustomed to evaluating personal writing and, for many, the standards for critique and discussion seem "slippery" in comparison with the more familiar criteria associated with orthodox scientific research reports. She suggests that as personal writing becomes more common among social scientists, researchers will need to develop new avenues of criticism and praise for such work. Different criteria from those applied to more traditional tales will be required to pass judgment on autoethnographies. The problem of criteria and how we might judge different tales will be focused on specifically in chapter 9.

Reflections

Autoethnographies and narratives of the self, when well crafted, have the potential to challenge disembodied ways of knowing and enhance empathetic forms of understanding by seeing our "actual worlds" more clearly. The stories are not just *about* the body; they are told *through* the body of the author. They come out of the body and are voiced in multiple ways that can connect people in their shared vulnerabilities, even though they may occupy different subject positions.

In reading the autoethnographies of Tsang (2000), Duncan (2000), and the others referenced in this chapter, we are taken into the intimate, embodied world of the other in a way that stimulates us to reflect on our own lives in relation to theirs.

According to Duquin, the memory work associated with autoethnographies is a particularly powerful method for investigating the importance of emotions in forming self-identity in sport and for examining the relationship between agency and social structure:

> Memory is tied to emotion; feelings make events significant. In memory work, replaying past emotions reveals the forces and everyday events that helped to shape self-identity. . . . Memory work reveals how emotions are socialized in sport and how individuals can become active agents in constructing their emotional lives. One major advantage of such methodology is that personal memory work exposes the complex interaction of various social statuses (for example, class, gender, sexuality) in the emotional patterning of individual lives.
>
> *Duquin, 2000, pp. 480-481*

Readers of autoethnographies may assume that the authors have not deliberately fabricated details and have limited themselves to events they "remember." As storytellers, however, the authors are concerned primarily with evocation rather than with "true" representation. In their texts, learning about the body, self, and identities integrates emotional and cognitive dimensions, and it emphasizes participating with rather than describing for the other. Here, it is interesting to note the view of Tinning (2000) in his review of *Talking Bodies: Men's Narratives of the Body and Sport* (Sparkes & Silvennoinen, 1999), which contained several autoethnographies.

> It is rather rare that an academic book actually "speaks" to me beyond the rational. It is rare to find one's right brain and emotions engaged in academic discourse. This edited collection is a book that

did these things for me. . . . At the outset I want to say that some of the stories in the book were (for me) very moving. Reading about Mark Sudwell's changing relationship with his father (an ex-physedor/gymnast with a crook back) made me think about violence, pain, frustration and the limitations of (his dad's) particular gendered/cultural construction of masculinity. This story took courage to write for it is deeply personal and revealing. . . . Martti Silvenoinnen's story of the accident of a young Finnish boy on the other side of the world (to me) sent a shiver down my spine. . . . Writing on "the secret life in the culture of thinness," Mikko uses his own personal diary kept over a period of about a year to give a moving account of his "becoming anorexic." . . . I was moved by his story and, equally important, I came better to understand the mired relationships between body appearance, self-love (and hate), food and exercise. I am glad he had the courage.

Tinning, 2000, pp. 90-91

In reviewing the same book, Skelton (2000) offers a story of his own about how a friend of his, a physical education teacher, has recently found out that he has ankylosing spondilitis (a form of arthritis that attacks the spine progressively, leading to chronic pain and loss of mobility). According to Skelton, the autoethnographies contained in the edited volume might begin to help his friend with some of the questions he is beginning to ask about the contradictory interrelationships between his body, sport, and his various identities. The narrative style of the autoethnographies is welcomed by Skelton for three reasons:

First, there is no simple closure with narrative—you tend to be left thinking aloud due to the ambiguities and complexities of the storyline. This supports further reflection and learning. Second, compared with many academic texts, narrative is very accessible. I read this book very quickly and I think this was because the stories were engaging and personal and theoretical sources were embedded within the account. Third, the narratives in the book offer a "critical" perspective on the relationship between men's bodies and sport, but one that recognizes multiple senses of self and shifting identities. Narrative therefore can offer a medium in which critical and postmodern sensitivities can be brought together in productive fashion.

Skelton, 2000, p. 28

The comments made by Skelton (2000) and Tinning (2000) do not devalue the use of abstract theory or deny its relevance for specific purposes in specific contexts. Nor do they suggest that we should never use stories as data. Researchers from a variety of disciplines are rightly interested in autobiographical tales both as a cultural product and as a social act, as well as a resource for investigating changing ideas about the body, self, and identity. Rather, the views expressed by Skelton and Tinning suggest that, as I have argued elsewhere (Sparkes, 2002), on occasion we need to consider repositioning our theories and data in relation to life as lived and be a little more cautious about how, when, and why we turn storied lives into categories and theories.

As Frank argues, "The grounding of theory must be the body's consciousness of itself. . . . Only on this grounding can theories put selves into bodies and bodies into societies" (1991, p. 91). In challenging the tyranny of abstraction, Frank suggests that bodies certainly do have problems among other bodies, "but the point is to hold onto the fundamental embodiment of these problems rather than allowing the problems to be abstracted from the needs, pains and desires of bodies" (p. 91). In this regard, autoethnographies can help us to hold onto the fundamental embodiment of problems and keep us closely connected to the needs, pains, and desires of bodies.

This potential is supported by Ellis and Bochner (1992). They point out that, rather than the usual privileging of cognitive knowing and the spectator theory of knowledge, in which knowing is equated exclusively with observing from a distance, authors of autoethnographies incorporate feelings and participatory experience as dimensions of knowing. Furthermore, by acknowledging the potential for optional readings, these authors give readers license to take part in an experience that can reveal to them not only how it was for the authors, but how it could be, or once was, for themselves as readers.

Autoethnographies, therefore, offer challenges to traditional ways of telling and knowing in sport and physical activity. For example, Ellis and Bochner (2000) note how autoethnography undermines the rational actor model of social performance. This is because the messiness and ambiguity of many autoethnographic texts tend to stress the journey rather than the destination, and therefore ellipses the scientific illusion of control and mastery that so often prevails in realist tales.

> The episodic portrayal of the ebb and flow of relationship experience dramatizes the motion of connected lives across the curve of time, and thus resists the standard practice of portraying social life and relationships as a snapshot. Evocative stories activate subjectivity and compel emotional response. They long to be used rather than analyzed; to be told and retold rather than theorized and settled; to offer lessons for further conversation rather than undebatable con-

clusions; and to substitute the companionship of intimate detail for the loneliness of abstracted facts.

Ellis & Bochner, 2000, p. 744

Of course, producing autoethnographies is not without its dilemmas. For example, the disclosure of hidden details of private life, inner feelings, and emotional experiences, especially in the academic world, is a risky business. As Innanen comments, "I often find myself thinking of the personal and professional risks involved in writing of this kind. Did I really want you as a reader to know about these things?" (1999, p. 131).

Besides the potential dangers of revealing intimate secrets about one's self, there are also the ethical dilemmas of writing about intimate others. For example, Ellis highlights the anxieties she experienced when writing about her own interaction as a daughter and caregiver with her aging mother:

As I write this final version, I wonder what my mother would say about the frame I've added now to my own story about her. Would she think it unnecessary, even interfering with the plot? Would she know what a frame is or why it is here? Should I feel obligated to read this last version to her? Did I write this story for my mother?

Ellis, 2001, p. 614

Thankfully, the mother likes the story, but the tensions of writing this kind of story remain. Ellis (2001) acknowledges that she has not yet resolved the larger ethical dilemma of writing about intimate others or the more particular issue of how much to tell about her own mother. With regard to the latter, Ellis notes how the emerging relationship she developed with her aging mother gave her more confidence in her decisions about what is appropriate to tell. Thus, the process of engagement with others is crucial. As Ellis points out, "You have to live the experience of doing research on the other, think it through, improvise, write and rewrite, anticipate and feel its consequences" (p. 615). Living the experience is a messy business, and Ellis acknowledges that, perhaps, her own attempts to answer the ethical questions in relation to her work only muddle, rather than clarify, what it means to ask and give permission to write about others. She emphasizes, "This muddle, however, is closer to the truth of my experience than a contrived clarity based on prescribed rules would be. And, for me, that's good enough for now" (p. 615).

As L. Richardson emphasizes, writing about one's life is not without its perils:

Writing about your life brings you to strange places; you might be uncomfortable about what you learn about yourself and others. You might find yourself confronting serious ethical issues. Can you write about your department without serious consequences to yourself and your students? What about your family? Who might be hurting? How do you balance "fact" and "fiction"? How do you write a "true" ethnography of your experiences? These questions, of course, are the ones that contemporary ethnographers ask themselves when they try and write up their "data" about other people. How different it feels when it is you and your world that you are writing about; how humbling and demanding. How up-front and personal-in-your-face become the ethical questions, the most important of all questions, I think.

L. Richardson, 2001a, pp. 37-38

By substituting the companionship of intimate detail for the loneliness of abstracted facts and by taking authors to places where they might feel uncomfortable, autoethnographies can be unsettling for those who produce them. In discussing dilemmas I encountered as a producer of autoethnographic tales (Sparkes, 2002), I noted how writing in this manner forced me to reflect on and recognize how, in my "normal" academic writing, I maintain tightly secured boundaries within me and beyond me, keeping various identities and selves separate, shored up, and protected from the swirling confusions I so often experience in my daily life. In this writing I tend to privilege rigor over imagination, intellect over feeling, theories over stories, and abstract ideas over concrete events.

Like Bochner (2001), I too was forced to acknowledge just how often I become a *divided self* in which fragments of me become distanced spectators to the lives of other people and to my own life, seeing but not embracing each other. As Bochner observes,

> The sad truth is that the academic self frequently is cut off from the ordinary experiential self. A life of theory can remove one from experience, make one feel unconnected. . . . Academic life is impersonal, not intimate. It provides a web of distractions. The web protects us from the invasion of helplessness, anxiety, and isolation we would feel if we faced the human condition honestly.
>
> *Bochner, 1997, p. 421*

For me, acknowledging the existence of this web of distractions and how it operates within me, on me, and through me to shape my life and the lives of those I am close to has not been an easy task. Untangling the web so that a different way of being in the world becomes possible is even harder. There are no easy solutions. As such, my own use of autoethnography can be seen as but one attempt to accept and nurture my own voice(s) and to acknowledge the multiple subjectivities and positions I inhabit, with a view to disentangling some strands from the web so that they can be reworked into a different configuration. In this sense, writing autoethnographically has provided me with a means to develop a greater sense of integration between the concerns that infuse the "private" and "public," or "academic," domains of my life. Whereas before I called on separate voices to engage with these divided worlds, suppressing one at the expense of the other, I now feel more willing and able to combine them in the same writing arena. Like L. Richardson (1992a), however, I do not know where this attempt at integration will take me.

Producing autoethnographic tales brings perils as well as pleasures, problems as well as possibilities and potentials. Certainly, many scholars in sport and physical activity might not be drawn to this genre as a way of exploring lived experience. Others, however, will connect more to this form of representation and sense the contribution it can make as a way of knowing that allows both the author and the reader to feel and understand differently the world in which we live together as embodied beings.

Chapter 6

Poetic Representations

If one of the goals of qualitative research is to retell lived experience and to make the worlds of others accessible to the reader, then some have argued that poetic representations are better able to achieve this goal than other forms of writing.

> The poet's business is to create the appearance of "experiences," the semblance of events lived and felt, and to organize them so [that] they constitute a purely and completely experienced reality, a piece of virtual life. . . . [poetry's] distinguishing marks, which make it quite different from any actual segment of life, is that the events . . . are simplified, and at the same time much more fully perceived and evaluated than the jumble of happenings in any person's actual history.
>
> Langer, 1953, p. 212

Calling on the spirit of the "poetic" described by Langer (1953), Ely, Vinz, Downing, and Anzul argue that poetry can be an evocative form of qualitative research communication since "Poetry allows for maximum input—in and

between the lines" (1997, p. 135). They also emphasize that "Different poems about the same data can help a researcher to rethink the data and to work on additional ways to highlight them. Creating poems from in-depth interviews has been an extremely successful activity for many qualitative researchers" (p. 136). Likewise, L. Richardson (2000) argues that "settling words together in new configurations lets us see, and *feel* the world in new dimensions. Poetry is thus a *practical* and *powerful* method for analyzing social worlds" (p. 933).

Of course, this is not to say that poetic representation is the only or best way to represent all social knowledge. Rather, as L. Richardson (2001b) suggests, for some kinds of knowledge, poetic representation may be preferable to representation in prose. She also claims "that poetic representation is a viable method for seeing beyond social scientific conventions and discursive practices, and therefore should be of interest to those concerned with epistemological issues and challenges" (p. 877).

L. Richardson (2000) also suggests that writing up interviews as poems honoring the speakers' pauses, repetitions, alliterations, narrative strategies, rhythms, and so on "may actually better represent the speaker than the practice of quoting snippets in prose" (p. 933). On this issue, Coffey and Atkinson (1996) point out that in all cultures and subcultures, there are characteristic rhetorical and even poetic forms. Accordingly, poetic representations can be used to point out and emphasize certain aspects of personal or cultural spoken style. For them, "one of the recurrent *analytical virtues* of stylistic variation and innovation is that they can help the textual representation to reflect aspects of the ethnopoetics of the everyday life under consideration" (p. 128).

Louisa May's Story

An early example that displays the features described is provided by L. Richardson (1992a). Dissatisfied with more standardized formats of reporting data and lived experience, Richardson took a 36-page transcription of a five-hour, in-depth interview she had conducted with Louisa May (a pseudonym), an unmarried, Southern, rural, Christian woman from a poor family, and fashioned it into a five-page poem titled *Louisa May's Story of Her Life*.

In constructing the poem about Louisa May's experiences, L. Richardson (1992a, 1997, 2000, 2001b) points out that, following social research protocol, she used *only* Louisa May's words, voice, tone, southern rhythms, and diction. The poem had to build on other poetic devices, such as repetition, pauses, meter, rhymes, and off-rhymes, to convey Louisa May's narrative. Thus, as L. Richardson (2001b) notes, "the words are hers, but the poetic representation, including the ordering of the material, are my own" (p. 883). To illustrate this point, she provides an extract from the beginning of her interview with Louisa May and invites the reader to contrast it with the poetic representation in its entirety. The extract from the interview follows, with Louisa May's many pauses noted as ellipses.

> Well, most important to say in . . . terms of that is that I grew up in the South—which puts a definite stamp on what you think you are and what you think . . . you're going to be.
>
> [Louisa May looks at the tape recorder] I remember my origins when I hear . . . myself on tape—that Lady Bird kind of accent—and I think, "Oh my Lord. I'm from Tennessee." I had no idea I sounded like . . . that. In any event that [being Southern] shapes it and in . . . terms of . . . aspirations. I grew up in a very poor . . . with parents . . . who were uneducated but who lived in a very normal sort of . . . middle-class neighborhood where we rented a house. So my . . . friends were not in the same situation, but no one ever . . . suggested to me that anything might happen *with* my life. So . . . when I was 12, I suppose, and with my friends—and they . . . really, ah very nice, wonderful friends, some of whom I . . . still see. I remember thinking at the time that I would . . . want a large number of children.
>
> *L. Richardson, 2001b, p. 883*

Because of limitations of space, I do not intend to include the poem in its entirety (for this, see L. Richardson 1992a, 1997, 2001b). Rather, I invite the reader to contrast the interview extract provided previously with the beginning of the poem, which follows:

The most important thing
to say is that
I grew up in the South.
Being Southern shapes
aspirations shapes
what you think you are
and what you think you're going to be.

> (*When I hear myself, my Ladybird
> kind of accent on tape. I think, "Oh Lord,
> You're from Tennessee."*)

No one ever suggested to me
that anything
might happen *with* my life.

I grew up poor in a rented house
in a very normal sort of way
on a very normal sort of street
with some very nice middle-class friends
   (*Some still to this day*)
and so I thought I'd have a lot of children.

L. Richardson, 2001b, p. 888

In moving from interview transcript to poetic representation, Richardson (2001b) has taken liberties with the placement of the words, but not with Louisa May's language or her sense-making process. In constructing the poem, Richardson also relied on the tape itself, which recorded Louisa May's accent, pauses, tempo, and asides—which are missing in the interview extract. Furthermore, Richardson emphasizes, because Louisa May returns again and again to the idea of "normal life" in her interview, she uses this as the central theme of the poem. Thus, L. Richardson features poetically what was of chief importance to Louisa May as she described her life.

Reflecting on the benefits of her poetic representation, L. Richardson (1992a) notes how she drew on both scientific and literary criteria as she tried to meet a desire to integrate the scientist and poet within herself. She signals further benefits: For example, she points out that, in the routine work of the sociological interviewer, the interviews are tape recorded, transcribed as prose, and then cut, pasted, edited, trimmed, smoothed, and snipped, "just as if it were a literary text, which it is, albeit usually without explicit acknowledgement or recognition of such by its sociological constructor" (1992b, p. 23). In contrast, she argues, "writing sociological interviews as poetry displays the role of the *prose trope* in constituting knowledge. When we read or hear poetry, we are continually nudged into recognizing that the text has been constructed" (1994, p. 522). Thus, the facticity of the *constructedness* is ever present for both the author and the reader.

Given that the poetic form also plays with connotative structures and literary devices to convey meaning, L. Richardson suggests that it commends itself to multiple and open readings in ways that straight sociological prose does not.

The poetic form of representation, therefore, has a greater likelihood of engaging readers in reflexive analyses of their own interpretive labors of my interpretive labors of Louisa May's interpretive labors. Knowledge is thus metaphored and experienced as prismatic, partial, and positional, rather than singular, total, and univocal.

L. Richardson, 1992b, p. 25

Furthermore, L. Richardson (1992b) suggests, poetic representations are more able than realist forms to work with interview transcription to recover the reflexive process though which the knowledge is created in interviews as *interactional* speech events created in particular contexts, which are themselves examples of lived experience. Poetic representations, therefore, are able to reveal the process of self-construction, the reflexive basis of self-knowledge, and the inconsistencies and contradictions of a life put together in speech as a meaningful whole for several reasons. First, she argues, a poem is a whole that makes sense of its parts; and a poem is parts that anticipate, shadow, and undergird the whole: "That is, poems can themselves be experienced as simultaneously whole and partial, text and subtext; the tail can be the dog" (p. 26). Second, she goes on to note that the experiencing person is a person in a body, and that poetry, because of the devices it consciously employs (for example, meter, cadence, and assonance), can recreate embodied speech in a way that standard sociological prose does not.

> Thus poetry, built as it is on speech as an *embodied* activity, touches both the cognitive and the sensory in the speaker and the listener. . . . Lived experience is lived in a body and poetic representation can touch us where we live, in our bodies. Thus, poetry gives us a greater chance of vicariously experiencing the self-reflexive and transformational process of self creation than do standard transcriptions.
>
> *L. Richardson, 1992b, p. 26*

Producing and sharing poetic representations can also have a range of benefits for the researcher-as-author. These include, L. Richardson (1992a) argues, a greater sense of personal integration and an enhanced ability to step into the shoes of the other, as well as into the other's body and psyche. As Richardson became more attuned to lived experiences as subjectively felt by the other, the experience affected her willingness to know herself and others in different ways, and it enabled her to see familiar sights in new ways.

Poetic Representations in the Social Sciences

The potential of poetic representations to act as a powerful and practical method for analyzing social worlds and to generate different ways of knowing about these worlds is evident in the work of social scientists who have chosen this genre (see Austin, 1996; Baff, 1997; Brady, 1998, 1999, 2000; Clough, 2000; Finley, 2000; Hones, 1998; Jones, 1999; Kiesinger, 1998; M.

Richardson, 1998, 1999; Schechter, 2001; Simonelli, 2000; B. Smith, 1999; P. Smith, 1999; Travisano, 1998, 1999; Weems, 2000).

For example, Austin (1996) opted to use a narrative poem to construct her experiences with Annie, another African woman, as they explored numerous issues through dialogue and conversation in the context of interactive research interviews. The narrative poem uses their spoken words, as well as the thoughts, memories, and judgments that emerged during and after their interactions, in a way that invites the reader into their lives and into the complex ways they attended to each other. Austin notes,

> As our conversations evolved, I discovered that my own eyes, personality, history, judgments, body, and sensibilities were integral parts of our interaction. It was important, therefore, to use a form of expression that would allow me to show these aspects of our experience being negotiated.
>
> *Austin, 1996, p. 207*

Kiesinger (1998) faced similar problems in her research using interactive interviewing to explore the emotional experiences of anorexic and bulimic women. She chose to construct evocative narratives in the form of poetic representations based on the detailed autobiographical accounts of the women involved. In part, this choice was fueled by Kiesinger's dissatisfaction with how standard forms of research writing in a variety of subdisciplines have portrayed eating disorders. She was particularly troubled by how rarely such accounts focus on the concrete lives and language of anorexic and bulimic women and on their own special ways of understanding and talking about their feelings, relationships, and experiences. For Kiesinger, the dry, clinical, and highly analytical nature of many "expert" accounts failed to convey the sense of how anorexic and bulimic women themselves understand and make sense of their lives and conditions. Furthermore, in these expert accounts, the voices of anorexic and bulimic women were strikingly absent.

Accordingly, Kiesinger (1998) drew on the lengthy life histories produced during the interactive interviews, as well as her own observations, field notes, and recollections about each woman, to compose an evocative narrative that expresses the ways the women involved experienced their conditions, understood their identities, and participated in close relationships.

> The evocative narrative as an alternative form of research reporting encourages researchers to transform collected materials into vivid, detailed accounts of lived experience that aim to show how lives are lived, understood and experienced. The goal of evocative narratives

are expressive rather than representational; the communicative significance of this form of research reporting lies in its potential to move readers into the world of others, allowing readers to experience these worlds in emotional, even bodily, ways.

Kiesinger, 1998, p. 129

Having conducted her PhD study on students' lack of engagement in language classes at a junior-senior high school in the United States, Baff (1997) chose to represent her findings in poetic form. For her, the poems were both the substance of the study and the lived experience of those involved in it.

They include many voices; those of the classroom teacher, the English teacher across the hall, the students, and myself can all be heard. I quoted directly from the discussion and interview transcripts wherever possible. I chose poetry not only because it is a natural form of expression for me personally but also because I wanted to give a three-dimensional picture of the experiences of all of us during the study. A narrative would have given too much linearity to a situation that appeared in my mind's eye as more circular.

Baff, 1997, p. 469

Echoing the sentiments of Coffey and Atkinson (1996), Ely et al. (1997), and L. Richardson (1992a, 1997, 2000), one of the main points emphasized by Baff is that the actual process of writing the poems was itself part of the analysis: "Trends appeared not only as I consciously thought about the data but also as I included various voices within one poem, and then another, and heard the same theme repeated across poems" (1997, p. 488). Baff also suggests that, for readers who have read many scholarly articles and books about English education, experiencing a literary or artistic representation of similar data may provide a different lens through which to view the same scenery. Therefore, for her, another viable role that poetic representations might have is to provide an experience that transcends any one particular study.

Baff (1997) makes the case that, for her particular study, poetry was an effective alternative method of representation because it enabled the reader to actually go through the same process in reading the results and analysis as the participants did during their literature class. Furthermore, Baff argues, using a poetic form of representation predisposes the reader toward a particular way of reading. For example, a reader who has prior experience and knowledge of poetry might expect to extract concentrated strong images and feeling from a poem and also to look for layers of meaning. "These expectations allow the reader to get deep into the data and interpretation in a nonlinear way, matching the many overlaid, connected layers of talking about literature in a secondary English class" (p. 488). Finally, speaking as a novice researcher, Baff argues for the reflexive use of poetry and other modes of artistic expression as a valuable tool among many for the researcher. Therefore, even in situations where the researcher chooses, or is required, to use a narrative or a standard research format, "working creatively along with traditional writing is a way to access the emic perspective of both researcher and participant. As researchers, we must beware of paradigm paralysis" (p. 489).

Reinforcing the points made previously is the research of Glesne (1997), which involved 10 hours of interview with Dona Juana, an elderly Puerto Rican researcher and educator. Of interest here is the strong case that Glesne makes for what she describes as poetic transcription. This involves the creation of poemlike compositions from the words of interviewees.

> Poetic transcription moves in the direction of poetry but is not neces-sarily poetry. . . . Poetic transcriptions approximate poetry though the concentrated language of the interviewee, shaped by the researcher to give pleasure and truth. But the truth may be a "small t" truth of description, re-presenting a perspective or experience of the inter-viewee, filtered through the researcher. It may not reach the large "T" truth of seeing "with the eyes of the spirit" for which poetry strives.
>
> *Glesne, 1997, p. 213*

In describing in detail her poetic transcription process, Glesne (1997) notes several rules that she imposed on herself. These included the rule that the words in the poetic transcriptions would be Dona Juana's, not hers: "I could pull Dona Juana's phrases from anywhere in the transcript and juxtapose them, and I had to keep enough of her words together to re-present her speaking rhythm, her way of saying things" (p. 205).

Prior to constructing six poetic transcriptions to represent Dona Juana's lived experiences, Glesne (1997) notes how the process began with the kind of coding and sorting that she usually undertook when dealing with qualitative data. That is, after reading and rereading interview transcripts, she generated major themes and then coded and sorted the text by those themes. Her desire to create varied portraits of Dona Juana helped to guide the development of the themes. Some were more data rich than others because Glesne had developed a series of interview questions specifically about these topics. Other themes contained less data because they emerged in the process of the interviews.

Having sorted the text in the manner described, Glesne (1997) felt ready to begin writing in a way that, for her, normally involved progressive coding, categorizing, and ordering of large data clumps. Instead, Glesne re-read all the transcripts under one theme and reflected, trying to understand the essence of what Dona Juana was saying so that it might be extracted and represented in a concrete, condensed form. Thus, Glesne began to write "using only Dona Juana's words to portray the essences that I understood" (p. 206). As part of this process, the previously used coding and categorizations were not totally abandoned, and some were useful in getting Glesne started on her task. Glesne recognized, however, that poetic transcription called for a different and less-ordered structure to analyzing the data.

According to Wolcott, *analysis* "refers quite specifically and narrowly to systematic procedures followed in order to identify essential features and relationships" (1994, p. 24). As a consequence, analytical writing in the form of realist tales tends to break up interview transcriptions and observation field notes into component parts, imposing a researcher-perceived order on events (see chapter 3). Glesne (1997) acknowledges that poetic transcriptions are also filtered though the researcher but that this involves word reduction while illuminating the wholeness and inter-connection of thoughts.

> Instead of piecing together aspects of Dona Juana's story into a chro-nological representational puzzle of her life (with pieces missing), I found myself, through poetic transcription, searching for the essence conveyed, the hues, the textures, and then drawing from all the por-tions of the interviews to juxtapose details into a somewhat abstract re-presentation. Somewhat like a photographer, who lets us know a person in a different way, I wanted the reader to come to know Dona Juana through very few words.
>
> *Glesne, 1997, p. 206*

Like Baff (1997), Kiesinger (1998), and L. Richardson (1992a, 1992b, 2000, 2001b), having gone through the process of producing a poetic transcription, Glesne (1997) comments on several benefits. First, this form of representation highlights the fact that it is the author who is staging and shaping the text, foregrounding the ethical dilemmas that thus arise. Glesne notes that poetic transcriptions create a third voice that is nei-ther the interviewee's nor the researcher's but a combination of both. Here, the researcher uses the interviewee's words to compose the pieces that tell a certain story, make a point, or evoke a feeling told, heard, and felt by either the researcher, the interviewee, or both. As a consequence, "poetic transcription disintegrates any notion of separation of observer and observed. These categories are conflated in an interpretive space"

(p. 215). Second, even though organizing a transcript into poetry or poetic transcription imposes a particular meaning on it, it can paradoxically "also 'pull out' meaning, moving the reader into the interpretive realm where the writer (and reader) make leaps, while staying close to the data" (p. 215). Finally, producing poetic transcriptions can allow researchers to see similar terrain (including themselves) in new and more complex ways.

Poetic Representations in Sport and Physical Activity

The potential benefits of poetic representation for both the author and the reader are evident in the work of those who have chosen to use this genre in sport and physical activity. Poets have often turned their attention to sport and the body. For example, a special issue of *Quest* in 1989 (Vol. 21, No. 2) was dedicated to poetry and art in sport and movement. The contents of this special issue illustrate how poets and artists are able to synthesize, assemble, and compose images that make connections to the subjective world of feelings and emotions.

More recently, four poems by Jackson (1999), which explore how his masculine senses of self were constructed via school sport and physical activity, appeared in an edited book on men's narratives of the body and sport (see Sparkes & Silvennoinen, 1999). Here is one of his poems.

Boxing Glove

My father couldn't stand a soft, moist palm.
Didn't want to sense his own grip slither
from who he thought he was.

He crammed my uncalloused skin
down into the blooded glove. My bunched,
wriggling fingers curled into pumped-up fist.

He couldn't stretch to his full height
until my skinny wrists were tightly
held in stiff leather.

His own treacherous longings—singing,
dancing, playing the piano like Charlie Kunz—
were hidden behind his hearty shoulder-thump.

As I soldiered on, through round after round,
pummelling my way out from the sneers,
my fighting hands froze into claws.

My fingers aren't mine any more.
I turn away from my father's searching hand.
Too proud to unclench a rock-hard fist.

Jackson, 1999, p. 48

Many women have also chosen poetry to communicate the significance of sport in their lives and the ways it can challenge gender oppression. For example, in the prelude of *Crossing Boundaries,* an international anthology of women's experiences in sport, the editors suggest that the poems and short stories it contains take us on a journey to the center of female identity:

It explores the complexities of female experiences as authors and athletes challenge cultural notions of gender, gender roles, and the very nature of femininity and sport. It is a chronology of attitudes, a catalog of experience, a survey of past and present voices—resigned, rebellious, detached and connected, searching and authoritative.

Bandy & Darden, 1999b, p. xiii

Other volumes of poems and short stories by women, such as those edited by Sandoz and Winans (1999) in *Whatever It Takes: Women on Women's Sport* and by Sandoz (1997) in *A Whole Other Ball Game,* make further contributions to an understanding of women's unique experiences in sport and physical activity.

In contrast, Swan (1999) includes a poetic representation in a larger ethnographic fiction (see chapter 8). This is based on his experiences of coming to understand the complexities, oppression, and ironies of masculinity via reflections on his own body and the bodies of other Australian men and boys in the spaces provided by changing rooms in sport centers. The poetic representation he produces draws on a study in an Australian Catholic secondary school on aspects of health promotion. As part of this study, an action research group involving year 7 to 12 students became interested in having cubicles in the boys changing room. Their interest in the cubicles arose in an effort to encourage higher levels of participation in school-based sport and physical education. It was school policy that the boys had to change for sport and physical education and had to shower afterward. The lack of privacy was seen as a factor affecting participation, and so the group agreed to see how significant this issue was to boys who didn't participate very much. Accordingly, using purposeful sampling six boys were interviewed.

Using the words from the interview transcript with one of the boys, called Bryan, Swan constructed in what he calls "loose verse format" a poem called *Changing (for) Bryan*. The words were instigated when Swan asked, "Bryan, could you tell me a little about your experiences in sport and physical education at school?"

Changing (for) Bryan

Well, because of how I'm perceived at school
It's most uncomfortable, you know, ridicule and teasing and stuff.

From year eight on, they'd call it 'jokes'.
"Why don't you come and get me poojammer,"
"Take a look at this one needle-dick."
"Oh you faggot, ah, perving are you."

They reckon I'm skinny, I act and can dance,
The clothes I wore were the same as theirs
You know, I hate ridicule and teasing and stuff.

I always worried about the change-room,
and, and, and (. . .)
what's going to happen in the showers?
Yuk, I don't want to do sport or PE.
Like, they make advances in
A dirty kind of ridicule way.

A hundred times I left my gear at home
Lots of notes from mum or from me!
I was sick in my stomach over sport and PE
I got a lot of detentions, some poor reports too.
It's the shower part that was really so bad,
"He's gonna fuck you, it's backs to the wall."

They'd spit and pee on me
I'm really weak when it comes to violence and that stuff.

I'd be saved if they teased Sam because of his weight
He's just laugh at the fat jokes, others too, because of their tool,
How far they'd matured
"No pubes, tree trunk, pin dick and things like that."
It's uncomfortable, ridicule, teasing and stuff.

You need to be able to do things that
Don't make you have to shower if you don't want to.
Teachers feel bad so they don't go in
People would grab me and simulate sex.
When I got older I could deal with it then.

In primary school I'd go home crying and hated school lots
At junior high I got blamed for not doing sport
But now it doesn't grab me at all
You know (. . .) the ridicule, teasing and stuff.

Swan, 1999, pp. 44-45

Talking about his poetic representation, Swan (1999) notes how the oppression of boys like Bryan strikes at us all since "Bryan as a metaphor for a marginalized version of masculinity suffers immensely" (p. 46). For Swan, his choice of poetry as a representational format allows both him and the reader to see familiar sites, like changing rooms, in new ways. Furthermore, Swan argues, poetic representation allows readers greater interpretive freedom to make their own sense of the events and people focused on.

I have tried to use forms of verse that allow the reader to feel the emotional context of the data and to produce a story. Clearly, poetry may better represent the speaker, whilst at the same time problematizing trustworthiness. There is no plot in this poem, as in narrative, yet voice is given to the reader to make sense, to feel and to rupture the tranquility of assumed relationships. *Changing (for) Bryan* arose from three pages of interview notes. Upon reading it Bryan expressed that he was very happy, that the words were very much his and he then said: "Make sure Endo and Jerker (two physical education teachers) read it won't you." He felt let down.

Swan, 1999, p. 46

Similar benefits are noted by Dowling Naess (1998), who utilized a variety of representational forms in her PhD thesis that explored the life histories of Norwegian physical education teachers. In one of her chapters, Dowling Naess offers a poetic transcription of the life story of one of her interviewees, called Randi, told during nine hours of interview. Details of Randi's life are made available in earlier chapters of the dissertation in the form of a realist tale. Dowling Naess chose to retell it in a later chapter as a poetic transcription, however, because she felt this form of analysis allowed for word reduction in a way that simultaneously illuminated the wholeness and interconnections of Randi's and the researcher's thoughts. This was important because, as Dowling Naess came to realize, there was an uncanny likeness between her life story and that of Randi, which forced her to reflect on how she herself had become trapped in teaching and how her own career had developed via series of fateful moments.

Furthermore, Dowling Naess felt the poetic transcription was the most appropriate medium (at that part of her dissertation) for exploring the complex personal and professional dilemmas experienced by Randi over time, and for displaying in an evocative manner the circular and contradictory nature of her career in teaching.

Teaching's Not for Me

You're expected to become something in my family.
A nurse, perhaps?
I like people.
Better still, a lawyer.
Keep it in the family.
Maintain respectability.

All I know is that teaching's not for me.
Never in a million years.
Remember Mrs. Weaks?
We ruined her life, gave her a breakdown.
How could we?
We were old enough to know better.

What are you good at, my form teacher asks?
Sport, I laugh! But, a career in sport?
Teaching, he replies.
No way!
Teaching's not for me.
Never in a million years.

He must be off his head!
Poor pay, no status, demanding pupils.
No, teaching's not for me.
Sports Studies.
Now, that does sound interesting.
Is that for me?

I'm good at sport and others think so, too.
A talent, some say.
I can coach as well.
Teaching's certainly not for me,
But PE and Sports Studies
Where's the harm in that?

This course is really cool!
My friends are green with envy.
Physical activity all day long
Few books.

No dank, musty library for me.
Although I still don't know what I want to be.

One year, two years and now it's almost three
A pity I don't know what I want to be.
I'm qualified to teach
But that's not for me.
I'm doing well in basketball, playing in the league
A supply post in school? Maybe.

I'm teaching now, just for a while.
It gives me time to decide what I want to be.
I'm independent,
Have a flat.
Teaching isn't as bad as all that.
Soon I'll know what I want to be.

I've given up studying law in the evenings.
Not enough time.
Not with the basketball,
And you've got to live!
There's a new curriculum now for PE
Maybe, a teacher, is what I want to be?

It's getting rather boring
Just teaching gym.
Is this it,
Life?
I mustn't get trapped.
What on earth shall I be?

I want to realise myself.
And teaching's not for me.
I like the kids
And the sport.
But what about the rest of me?
. . . the rest of me?

A woman in the eighties can do anything
A scientist, an engineer?
It can be done!
But who's that guy?
That one, there.
I'd like, him, to get to know me.

He's asked me to marry him,
I've never been so happy.
A wife,
A home.
I'll teach a little longer
And then decide what to be.

I'm teaching Sports Studies now
At a different school.
It's challenging,
Theoretical.
I can use my education.
Is this what I want to be?

I'm having a break from teaching
I'm a mother now.
To David,
Our first born.
How precious life is,
It's wonderful to be.

Experience and maturity enhance my teaching,
I do my best for pupils.
Not worth the trouble?
Who said that?
Am I not worth the trouble?
You're not simply saying something about PE.

Once again, I'm not sure what I want to be.
A teacher, a joker?
That's not for me.
Respect.
I deserve respect.
What can I be?

I want to jump off the carousell,
To start again.
Business administration?
Law, like Mum?
It's never too late!
But, teaching's not for me.

I'll delay a while, the family's grown.
And then, there's that damn loan!
A housewife?
Not for me.
Is the grass greener?
I'll wait and see.

I know the ropes, or thought I did.
Not more paperwork!
Another meeting,
Plan?
Whatever happened to teaching?
Oh, what do I want to be?

The childminder's ill,
And he has an important meeting.
I'll stay home,
'Hurrah!'
David and Anne jump for joy,
Is this how we wanted life to be?

Time, used to be measured by the clock,
Now it's just something I don't possess.
No time for family,
Nor for work.
And it's running out fast,
What shall I be?

Leisure pedagogy might hold the key,
Twenty credits is all I need.
Nine to five,
And weekends free!
It's going to be tough,
But family life's for me.

There's no one to turn to
At school for advice.
I'm frightened,
Bewildered.
What shall I be?
Teaching can't be the last stop for me.

Dowling Naess, 1998, pp. 226-230

Reflections

Brendan Kennelly is professor of modern literature at Trinity College in Dublin and an internationally renowned poet. Reflecting on a collection of his early poems in a book titled *Breathing Spaces*, he makes the following observation.

Poetry is an opening of the doors of rooms that are never fully known; the poet is an eternal door-opener while at the same time living with the sense of always being on the outside, of not being entirely at home even where he might be said to belong. For me, poetry is an entering into the lives of things and people, dreams and events, images and mindtides. This passion for "entering into" is, I believe, the peculiar vitality of the imagination.

Kennelly, 1992, p. 10

As described by Kennelly (1992), the quest of the poet would appear to have much in common with that of the qualitative researcher. It would therefore seem sensible for researchers to harness the power of poetry and use it as a resource in representing their work. When well crafted, poetic representation can engage the reader emotionally; reflect the ethnopoetics of everyday life and use the individual's voices in sensitive and meaningful ways; touch us where we live, in our bodies; allow the researcher and reader to step into the shoes of the other; and see, feel, and analyze the familiar in new ways. As Coffey and Atkinson (1996) note regarding L. Richardson's (1992a) poem of Louisa May, "The effect is striking and has—for some readers at any rate—an emotional force, coupled with a sense of how Louisa May constructs her life through telling it, that might not come through a more prosaic account" (p. 128).

Poetic representations, then, under certain circumstances, can have the potential to generate understanding in a way that is different from more traditional forms of representation. As L. Richardson (2001b) notes, "Poetic representation, I submit, is a practical and powerful, indeed transforming method for understanding the social, altering the self, and invigorating the research community that claims knowledge of our lives" (p. 888).

For some, however, the association of poetic representations with poetry might seem a little intimidating. Some might feel that they do not have the confidence or the necessary literary skills to attempt representing their data in poetic form. With beginners in mind, L. Richardson (2001b, pp. 881-883) shares recommendations and practices that she and others have found useful in constructing poetic representations from interviews and interview experiences. These include the following.

- Take a class in poetry, attend poetry workshops and poetry readings, join a poetry circle, and read contemporary poetry.
- Remind yourself:
  A line
  break does
  not
  a poem
  make.
- Revise, revise, revise. Read the poem aloud and get other people to read it aloud. Put the poem away for a while. And then revise some more. Write different poems about the material.
- Do not imagine that all poetic transcriptions will be publishable.
- Do not imagine that your work cannot be published. Nor need you imagine, if you are a student, that your adviser will not approve your work.
- Because poems (and prose, too) are more interesting if they include metaphoric language, construct your interview schedule in such a way as to elicit images and similes. Implement this approach early in the research process.

One does not have to aspire to be a professional poet to produce poetic representations of data from qualitative research projects. Knowing this—and not feeling the need to compare ourselves, should we attempt such a task, to the likes of Brendan Kennelly—can make the enterprise less intimidating. Here, we can take heart from the reflections of Woods (1999) on his own poetic representation of interview data with a teacher about her experiences of a government inspection of the school she worked in. His poem, called "Ofsted Blues," is 25 lines in length and uses the teacher's own words in an attempt to present the essence of her response in a way that creates a vivid, immediate, emotional experience for the reader, so as to integrate the sociological and the poetic at the professional, political, and personal levels.

For Woods (1999), this piece was an addition to, rather than a replacement for, a more traditional analysis. That is, it offered another dimension to his understanding of the research problem. As part of a larger project, the interview on which the poem is based was analyzed with others for common themes and categories, and some of these were starkly revealed in the poem. Woods acknowledges his weaknesses as a poet.

"Ofsted Blues" will not make the 20th century book of verse, but I found it a useful way of getting these points over on an overhead projector during presentations, and emphasizing the prominent features of the experience encapsulated all in one display. In neither case would it have been possible to present Shula's original unedited utterance, which ran to several pages. The constraint here is not just one of the publisher's restrictions on word count, though that is a serious consideration; but also one of judging how best to get the teacher's feelings over to readers or audience. I make no claim to one presentational form being superior to the others. They are shaped for different situations and different audiences.

Woods, 1999, p. 59

Coffey and Atkinson (1996) reinforce the view that the use of poetic representations needs to have an analytic purpose. For them, there is a danger in such exercises of producing emotional or aesthetic effects simply for the sake of producing them. In this situation, Coffey and Atkinson suggest, they can appeal to "inappropriately self-indulgent displays of cleverness on the part of the author" (p. 129). They go on to emphasize that there may be occasions when the "ethnopoetics of a given culture may be served better by a faithful rendition of *their* forms, rather than the imposition of an author's aesthetic judgments" (p. 129).

The data generated in qualitative research do not self-evidently lend themselves to the construction of poetic forms, and even when they do, such a transformation may not be the best choice in all circumstances. As Glesne notes,

It depends on the inclination of the presenter, the nature of the data, the intended purpose for writing up one's research, and the intended audience. . . . I would not argue that all data be written up as poetic transcription any more than I'd advocate always using matrices to convey analytical patterns.

Glesne, 1997, p. 218

Therefore, both poetic transcription and matrices can be effective ways to analyze and re-present data; it's just that some experiences may be best expressed in a particular form on certain occasions.

For some, poetic representations may be the preferred way to explore certain dimensions of sport and physical activity. Scholars in this domain will need to make strategic choices regarding if, when, and how they craft their data into a poetic form. This form of representation (like all the others considered in this book) is not appropriate to all situations and all audiences. When used wisely, however, poetic representations are a powerful means of understanding phenomena in new and exciting ways.

Chapter 7

Ethnodrama

According to Coffey and Atkinson (1996), "The idea of ethnodrama is to transform data (dialogue, transcripts, etc.) into theatrical scripts and performance pieces" (p. 126). There is a long tradition of using drama in educational, therapeutic, and social change contexts, with some projects employing participatory strategies in script development and using the actual voices of members of the community under study. Although research has certainly been a part of this work, the emphasis has for the most part been on political, educational, or aesthetic considerations, or a combination thereof. Drama that emphasizes research is a much more recent phenomenon, with only a handful of authors publishing their attempts to foreground research in the construction of drama. Indeed, McCall (2000) notes that it was not until the late 1980s and early 1990s that sociologists began to turn their ethnographic field notes into performances and that theatre artists and academics in performance studies began to produce or adapt ethnographies in order to perform them.

As if heeding the call by Denzin (2000) to "seek a set of writing practices that turn notes from the field into texts that are performed" (p. 903), scholars in the social sciences have now begun to explore a diverse range of topics via dramatic productions. These have been variously labeled ethnographic drama or ethnodrama (Ellis & Bochner, 1992; Mienczakowski, 1995, 1996, 2001; L. Richardson, 1993; L. Richardson & Lockridge, 1991; Wellin, 1996), performance texts (Denzin, 2000; Paget, 1990; Pifer, 1999), performance ethnography (McCall, 2000), readers theatre (Adams et al., 1998; Donmoyer & Donmoyer, 1995, 1998), research-based theatre (Gray, 2000a; Gray, Ivonoffski & Sinding, 2002; Gray, Sinding, & Fitch, 2001; Gray, Sinding, Ivonoffski, et al., 2000), dramatically scripted narratives (M. Miller, 1998), and playwriting as critical ethnography (Goldstein, 2001).

Despite the different terminology used, Denzin (1997) suggests that the performance text can take one of the following forms, and that these forms must be distinguished from "staged readings" in which one or more persons hold a script, the text, and read from it. These forms are as follows:

> Dramatic texts, such as rituals, poems, and plays meant to be performed; natural texts—transcriptions of everyday conversations turned into natural performances; performance science texts—fieldwork notes and interviews turned into performance texts; and improvisational, critical ethnodramas that merge natural script dialogues with dramatized scenes and the use of composite characters.
>
> *Denzin, 1997, p. 99*

Often these kinds of production draw on data generated as part of research projects or call on the personal experiences of the researcher/author. For example, Ellis and Bochner (1992) draw on their shared experience of abortion to present a personal account narrated in both the female and male voices. In contrast, others have developed ethnodramas based on data generated via more conventional means, such as interviews, observations, documentary sources, and prolonged ethnographic fieldwork. For example, Gray, Sinding, Ivonoffski, et al. (2000) call up the material from an interview-based study about the information needs of women living with metastatic disease and the perspectives of medical oncologists about those issues. Likewise, Mienczakowski (1996) notes that his work on the experiences of schizophrenic illness and persons undergoing detoxification processes involved prolonged, intensive participant observation over four months in health settings, coupled with open-ended interviews (see also Donmoyer & Donmoyer, 1995; Goldstein, 2001; M. Miller, 1998; Pifer, 1999; Wellin, 1996).

Capturing Lived Experience

Why then transform data and analysis based on interviews, observations, participant observations, diaries, and so on into a dramatic production? For many of the authors mentioned above, this transformation was necessary to include the experiences of those who participated in the study (including the researcher-author) in ways that remained true(r) to those experiences. More conventional forms of reporting are not seen as capable of accomplishing this task. For example, Denzin argues, "Performance text is the single, most powerful way for ethnography to recover yet interrogate the meanings of lived experience" (1997, p. 94). Pifer also notes, "I believe that through performance the lives, voices, and events presented will have a life and power not possible through other forms of representation" (1999, p. 542). L. Richardson supports these views when she comments:

> Drama is a way of shaping an experience without losing the experience. . . . it can reconstruct the "sense" of an event from multiple "as-lived" perspectives; it can allow all the conflicting voices to be heard, relieving the researcher of having to be judge and arbiter; and it can give voice to what is unspoken. . . . When the material to be displayed is intractable, unruly, multisited, and emotionally laden, drama is more likely to recapture the experience than is standard writing.
>
> *L. Richardson, 2000, p. 934*

Reflecting on her own use of playwriting to represent the experiences of four years of critical ethnographic fieldwork in a Canadian multilingual high school, Goldstein (2001) notes how the performance of her playwriting challenges fixed, unchanging ethnographic representations of the research subjects. She emphasizes that the performed ethnography provides her with a way to breathe life into her data and present multicultural issues from a variety of student standpoints, allowing her characters'/subjects' voices to mingle in polyphonic conversation. Goldstein also points out that performed ethnography "offers opportunities for both comment and speechlessness. Finding a way to represent silence in ethnographic writing is difficult. Yet, there were many students in my study whose silence at school needed to be heard" (p. 296).

Against this backdrop, Ellis and Bochner (1992) offer the ethnographic drama about their abortion experiences as a form of narrating personal experience in which they are the experimental subjects and their own experiences are treated as primary data. The goal is to lead readers through a journey in which they come away with a sense of what the experience must have felt like. Recognizing that the literature rarely reflects the meanings and feelings embodied by the human side of abortion, Ellis and Bochner wanted to tell their story in a way that would avoid the risks of dissolving the lived experience in a solution of impersonal concepts and abstract theoretical schemes.

> We have tried to be faithful to our experience, but we understand that the order and wholeness we have brought to it through the narrative form is different than the disjointed and fragmented sense we had of it while it took place. Perhaps this is the way in which narrative constitutes an active and reflexive form of inquiry. Narratives express the values of narrators, who also construct, formulate and remake these values. A personal narrative, then, can be viewed as an "experience of the experience" intended to inquire about its possible meanings and values in a way that rides the active currents of lived experience without fixing them once and for all.
>
> *Ellis & Bochner, 1992, p. 98*

Reflecting on the feelings that their work might evoke in the reader, Ellis and Bochner (1992) recognize that while identification and empathy are likely, these reactions are not the only nor necessarily the most desirable ones. They hope to reveal not only how the experience was for the authors but also how it could be, or may once have been, for the readers. As a consequence, readers are made aware of similarities and differences between their world and that of the authors. It therefore becomes possible for them to see the other in themselves or themselves in the other.

Talking of their work with eighth-grade students, Donmoyer and Donmoyer note that because they wanted to be as true as possible to what they believed were the intentions of the students, "A readers theatre script seemed a better way . . . than a social science research report" (1995, p. 408). Likewise, M. Miller (1998) felt that traditional forms of reporting would not allow her to create multivocality, so instead she chose dramatically scripted narratives as a way of reporting what she had come to understand about her participants' understandings and about her own process as a researcher.

> Dramatically scripted narratives provide possibilities for the layering and interweaving of voices, theories, statistics, and accounts of the research process. Dramatic scripts can (re)present the experience of engaging in research activities and "attach significance" to the "data." . . . Thus, performance of dramatically scripted research not only illustrates the research process but takes on the form of a powerful embodiment of the data.
>
> *M. Miller, 1998, p. 69*

Similarly, having conducted ethnographic fieldwork in a residential group home for older women diagnosed with Alzheimer's disease, which revealed how the coercive dynamics of institutional life were less a function of bureaucratic scale than of medical ideologies and exploitative labor practices that produce "commodified" regimes of care, Wellin (1996) opted to create an ethnodrama. For him, "performance was a vehicle for enacting these coercive relations, and their ongoing impact on residents" (p. 500). As he explains,

> The resources of performance allowed us to express—more fully than through writing—the embodied and emotionally-charged nature of interaction in the setting. Rather than isolate themes, such as objectification or narratives of self-hood, we were better able to convey the dynamic interplay of these processes. We could present simultaneous speakers, para-linguistic information, and nuances of physical contact—along with the surface meanings of speech. And, because of the ostensibly domestic scene (a living room) and close proximity between the residents, their accomplished inattention to one another was unsettling for audience members. In sum, the performance aesthetic is more amenable to experiential ambiguities than has generally been true in the tradition of academic sociology.
>
> *Wellin, 1996, p. 508*

In combination, the authors mentioned in this section make a strong case for the use of ethnodrama as a means of connecting with the lived experience of others and ourselves in a research context.

Reaching Wider Audiences

For the scholars previously mentioned, transforming data into dramatic productions has been a way of getting findings across in a thought-provoking, engaging, and accessible manner to a diverse audience—an outcome that is preferable to the fate of many manuscripts, which lie unread, or at best skimmed over, on library shelves, or are commented on occasionally by other academics. For example, Goldstein (2001) wanted her ethnographic writing on multicultural issues to engage teachers as active participants in critical analysis so that they could generate meaningful schooling. To achieve this goal, she knew her findings and analyses needed to speak *to* teachers rather than *at* them. Therefore, Goldstein presented her findings in a play titled *Hong Kong, Canada*. Based on her experiences of presenting the play to a variety of different audiences, Goldstein acknowledges that performed ethnography "has the power to reach larger audiences outside my classroom and encourage reflexive insight into the experiences of schooling in multilingual / multiracial communities" (pp. 296-297).

For Mienczakowski (2001), performed ethnography may provide more accessible and clearer public explanations of research than is frequently the case with traditional, written report texts. To create community understanding and an awareness of mental health and alcohol and drug issues, Mienczakowski (1996) created two plays, *Syncing Out Loud* and *Busting*, from his research data. Performance venues were carefully chosen to guarantee that the plays reached their specific target audiences. For example, by researching in institutional settings and performing the plays in or close to those settings, he guaranteed that audiences would include health professionals, caregivers, and care agencies. Representatives from major government health agencies and from other teaching institutions were also invited to special forum presentations, and wide media coverage further ensured public and professional interest.

As the cast consisted of large numbers of nursing and theatre students, peer-group support for the productions was never in doubt. Similarly, schools were invited free to performances to encourage high school student attendance. The final validating (invited) audience for *Busting*, for example, consisted of more than 400 persons, including general practitioners, human resource officers from large multinational corporations, psychiatric nurses, students, health consumers, psychiatrists, government agency representatives, academics, health area managers, health administrators, informants and their families, police representatives, and interested others (Mienczakowski, 1996).

As Gray, Sinding, Ivonoffski, et al. (2000) argue, by reaching a wider audience and providing insights via their engagement in dramatic material, these presentations enhance the potential for institutional and individual change. In this sense, researchers might be better able to take on their often-neglected responsibility to have their work make a difference in the everyday world. For example, Mienczakowski (1996), talking of reactions to *Busting*, notes how the informants, in particular, felt that at last they were being listened to and heard by groups who often ignored them. Likewise, many of the nursing students working on the project expressed a profound change in their understanding of persons involved in these health issues. Others were also affected:

> In the forum sessions for *Busting*, a senior corporate human resources officer for a large multinational corporation (specifically invited to the performance) expressed a newfound sympathy and understanding for dealing with issues of alcohol abuse in the workplace. Her intention at the end of the forum session was to review her dealings

with alcohol-affected employees and introduce drug and alcohol awareness education sessions into the workplace. Several months after the performance she had instigated changes in the company's employee health awareness program and was seeking further funding to assist in counseling programs for alcohol-affected employees.

Mienczakowski, 1996, p. 259

Goldstein expresses similar views with regard to the impact of her play, *Hong Kong, Canada:*

When I am very lucky, the audiences and performers . . . leave the room and the auditorium changed in some way. Sometimes students tell me that the work we have done with *Hong Kong, Canada* has helped them question or rethink their own teaching practices. It is at these moments I know the play has been persuasive and has facilitated questioning of social reality. The potential of performed ethnography to support critical teacher education seems particularly rich.

Goldstein, 2001, p. 297

Ethnodrama is not intended to be performed solely for small audiences of like-minded academics. If such a situation were to prevail, then ethnodrama might rightly be characterized as just a newish fad in representation, and questions could certainly be raised about its increased value. The critical point seems to be the need to move the influence of research outward from universities and other academic settings to include the communities that have been originally studied, the general public, and other interested and influential audiences.

Truer to Life?

According to Goldstein (2001), presenting ethnographic data in the form of a play allows the participants involved in the research and other people to watch the performance with a view to ratifying or critiquing its analysis. Likewise, Gray, Sinding, Ivonoffski, et al. (2000) also show a concern for validity issues when they argue for research-based theatre that attracts those with a predominantly realist (as opposed to constructivist or postmodern) perspective. The suggestion is that it has advantages over purely textual reports in terms of validity (i.e., remaining true to qualitative research data and ultimately to lived reality).

This is because it sustains connections to bodies, emotions and the full range of sensory experience that was present in the original data gathering situation. Audiences, often comprised of [sic] members of groups under study, can further validate research-based drama through provision of post-performance feedback, potentially affecting the shape of future presentations.

Gray, Sinding, Ivonoffski, et al., 2000, p. 138

For their production of *Handle With Care? Women Living With Metastatic Cancer*, Gray, Sinding, Ivonoffski, et al. (2000) created a partnership with a local theatre group for older adults (see also Gray, Ivonoffski, & Sinding, 2002; and Ivonoffski & Gray, 1999). As such, the group working on the production comprised researchers, women with breast cancer (two with metastatic disease and several who remained well after the treatment), and actors who were part of the theatre school.

The research team provided background information. Women with breast cancer told their personal experiences. Actors responded to what they heard and raised their impressions, fears and concerns. Transcripts from the studies with breast cancer patients and oncologists were circulated among the groups, and subsequently discussed.

Gray, Sinding, Ivonoffski, et al., 2000, p. 139

In the scripting of the play, the intention was to represent the perspectives of women with metastatic breast cancer in a way that they would recognize. This required a strong reliance on the interview transcripts to guide the writing of the script.

Continually, we returned to the words that women had spoken during the original focus group discussions. But we also looked for guidance from the two women in our project with metastatic disease. They were able to provide feedback when content or tone began to stray from the realities of their experiences. Although this same correcting influence was theoretically available through referral to transcripts, the immediacy and intimacy of women living with the disease was exceedingly compelling for limiting excesses in the expression of artistic license, or departures into intriguing but unessential byways.

Gray, Sinding, Ivonoffski, et al., 2000, pp. 139-140

This is not to say that accuracy of data overshadowed the artistic requirements of staging a play. Gray, Sinding, Ivonoffski, et al. (2000) note that while they felt it important to stay closely connected to the research base of the studies, they collectively agreed to numerous explorations beyond word-for-word excerpts from transcripts. Some of these departures were simply to allow for clearer expression of thoughts that study participants had articulated. In contrast, "Other explorations were more related to artistic considerations, and a desire to produce a compelling presentation that would entertain and engage the audience, not just inform them" (p. 140).

Similarly, in the scripting sessions in *Busting*, Mienczakowski (1996) notes that there was an adherence to, and preference for, verbatim account work as opposed to fictionalized account work. In particular, the informants in this study felt that unnecessary literary and plot fabrication would render the performance merely "a fiction" as opposed to "a truth." Mienczakowski points out that the counterargument of fictional accounts appearing true, and in many cases appearing more true than verbatim account work, was presented to the informants. They dismissed this view because they felt that the project had credence "only if the audience tacitly understood that the play's authority rested on its factual research status and could not be dismissed as authorial invention" (p. 248).

Of course, as Mienczakowski (1996) notes, where necessary fictional inclusions were inserted into the *Busting* script to link plot, subplot, and narrative. These inclusions were fictitious in the sense that they were amalgamations of separate elements of informant data and were not direct verbatim transcriptions of single interviews. Mienczakowski points out that most often they included scenarios constructed by informants as typical interactions with clients or health professionals in the domain under study. Therefore, "Fictional links were always based on informant accounts or anecdotes and were considered plausible by informant. When no such consensus was achieved, scenes were deleted from the text" (p. 248). This negotiation and attention to accuracy was seen as crucial in a process that aspired to give disempowered health consumers a voice in the community.

> To recontextualize and reconstruct their words unnecessarily and artificially to appease the aesthetic conventions of academic and literary traditions would have been to reduce further the significance of the voices of the informants and thereby act to disempower them.
>
> *Mienczakowski, 1995, p. 363*

The validation process was ongoing and dynamic. For example, at the rehearsals of the play, informants were involved to confirm or refute that the physical and semiotic representation of their ethnographic realities were authentic and cogent. Thus, informants participated in rehearsals and guided the actors. Likewise, actors undertook periods of immersion in actual clinical settings. Furthermore, all the rehearsals involved a range of health representatives, and the stage setting was constructed so that nursing and health informant behaviors, routines, and so forth, could be portrayed accurately.

For Mienczakowski (1996), in constructing authenticated accounts of informant experiences, the ethnodrama process relies on the notion of *vraisemblance* as being something more than just the creation of plausible accounts of informants' lives. *Verisimilitude* may be said to depict a similarity with a given truth. In contrast, vraisemblance involves a notion of probability or likelihood. Therefore, Mienczakowski argues, since the ethnodrama seeks to depict only what is "given as truth" or authenticated as "probably true" by informants, "it uses vraisemblance to describe its scripted content and messages as being 'of' or 'from truths' rather than 'similar to truths'" (p. 258).

In terms of vraisemblance, it is interesting to gain audience feedback. After performances of *Handle With Care?* both to health professionals and to the general public, Gray, Sinding, Ivonoffski, et al. (2000) distributed questionnaires. From seven Ontario cities, attendees at public performances returned 507 questionnaires (estimated at 60 to 70% of attendees). Many attending the public performances were themselves cancer patients (41%), were family members or friends of patients (48%), or were health care professionals or volunteers working in a health care context (48%). (Audience members checked as many of these categories as applied.)

According to Gray, Sinding, Ivonoffski, et al. (2000), all (100%) of those attending presentations for the public agreed or strongly agreed that they enjoyed the drama and that they benefited from seeing it. Similarly, 99% agreed or strongly agreed that there "was a lot of truth in this drama" (p. 140). The vast majority (99%) also agreed that the fact that the drama was based on research made it more "true to life" (p. 140). Most (96%) expressed a desire to see other dramatic productions about living with cancer.

A total of 249 feedback forms (estimated at 40 to 50% of attendees) were gathered from health professional audiences in eight Ontario cities. All (100%) respondees agreed or strongly agreed that they enjoyed the presentation and that the format of the presentation was engaging. All attendees also agreed or strongly agreed that the use of research transcripts to create the presentation "increased its validity substantially" (p. 141). Most (95%) were in agreement that dramatic presentations like *Handle With Care?* do have a place in hospital rounds. Most also felt that the issues presented were relevant to them (95%) and useful for thinking about their clinical practice (93%).

Feedback we received from both health professionals and public audiences revealed that the research foundation for *Handle With Care?* was important for ensuring a sense of relevance. While theatre can certainly be profound without a research base, audiences that are orientated towards empiricism (e.g. health professionals) appear to be more receptive and comfortable with "data" that have been accumulated within traditions of inquiry that they respect.

Gray, Sinding, Ivonoffski, et al., 2000, p. 142

In summary, claims have been made that ethnographic drama has not only the potential but the proven ability to give voice to what may be unspoken, and to give voice more accurately to those who consider themselves without power. It is also better able to represent lived experience, from multiple and contested perspectives, to a much wider audience than other forms of representation. Furthermore, ethnodrama does this in ways that are more authentic, evocative, and engaging. This being the case, it is surprising that scholars in sport and physical activity have made little use of this genre.

A Lonely Exemplar in Sport and Physical Activity

The use of ethnographic drama in sport and physical activity is unusual. A rare exception is the work of Brown (1998), who looks at the world of physical education teacher education (PETE) students. For Brown, the purpose of her play, *Boys' Training,* is to provide a "creative but empirically grounded" (p. 84) insight into the negotiation, social positioning, and identity construction process that occurs for many young males entering into PETE. It attempts to highlight issues relating to hegemonic masculinity in PETE and, in particular, to illuminate the processes by which certain values and characteristics are celebrated and becoming an "in" male physical educator is legitimized and reproduced.

In the tradition of research-based drama discussed earlier in this chapter, Brown (1998) points out that the situation and story portrayed in the play is based on a traditional social function and built around the stories of the young men who participated in it.

It is important to note that while the story has been fictionalized, the description of activities is based on a number of data sources including interviews, observations, and photographic representations. The language used in the construction of the play has been drawn from students' own descriptions of the event. Direct quotations from interview transcripts are presented in italics.

Brown, 1998, p. 84

The play also includes a series of brief narrations that, according to Brown (1998), set the scene and provide an interrupted reading where certain aspects of the play and the boys' activities are explained in more detail. This tactic is similar to that used by Pifer (1999), who included himself in the script as a narrator and who punctuated the scenes in his play on race with the words of various theorists to provide a critical perspective.

Boys' Training *by Leann Brown (1998, pp. 85-90)*

Characters

Mick Davidson: Davo to his mates. Mick came to university full of stories and expectations about living the high life. He planned to party hard at university before "settling down" into a career of physical education teaching. Socially it was better than he expected. It didn't take him long to earn the title of "legend" among his peers and he has spent the last three years trying to maintain this status. This is Mick's final year at UBU (You Beaut Uni) and he and his mates are planning to make it a "big one." His mates will tell you that he is a great all-round sportsman and is a "bit of a leader." The one thing he has been looking forward to this year is taking over the mantle as one of the fourth year "boys" and "showing the ropes," so to speak, to the first years.

Paul Simpson: Paul was absolutely wrapped at getting into P.E. at UBU. He'd struggled during year twelve and wasn't sure he would make it into the course with his tertiary entrance score. Ever since finding out that he had been accepted, he hit the party scene hard. "O" week became a never ending celebration for Paul and he had really started to make a name for himself within the first year P.E. group. He desperately wanted to be popular and pulled out all the stops at trying to fit in with the "P.E." image by wearing the right brand gear, attending all the social functions and even actively seeking out other "popular" boys to hang out with. Paul believes that the hardest part of being a "phys edder" is behind him, getting accepted into the course, and that he can now juggle his study load and still go out, have a good time and party hard. His new uni mates have made it quite clear that socializing is "expected" if you are to be one of the "boys."

Act one

Narrator: Guys and girls night has arrived. A night with a long history in the UBU physical education social calendar. The girls are partying at a venue a few streets away from the "main" venue, the

Eureka Arms, where the boys are eagerly awaiting the females' arrival later in the night. It's about mid-way through the evening and Davo and his mates have gone in hard, just to show the first years what it's all about. Their efforts at planning activities, entertainment and drinking games has had the desired effect and everyone is having a great time. Davo is happy, his "boys" are partying and the first years, nervous and keen to be a part of it all, are getting pissed. The funnels are about to begin and Davo spots a young "JAFFY" (Just Another Fucking First Year) who he thought was a bit of a "go-er."

Davo: (Moves away from the stage to top up his glasses from jug. Looks up and spots Paul laughing and sculling drinks with his mates). **(Voice over):** There's the JAFFY who puked over the table at the boat races in "O" week. (Laughs to himself.) Looks like he's giving it a good dip again tonight. SHIT, he's sculling now. What's he going to be like by the time the girls get here. (Laughs to himself and shakes his head). A DEAD loss but . . . he's got potential. I reckon we ought to get him on stage. See what he can do. (Finishes filling his glass, moves towards Paul and grabs him firmly by the crotch.)

Davo: How they hangin'.

Paul: **(Voice over):** What the . . . I know this guy, he's the fourth year that everyone talks about. A real party animal. What's he want with me? (Paul, surprised at the approach. Steps back, laughs and takes down a long slow draught of beer.)

Davo: You're giving it a real dip then, hey?

Paul: Yeah (trying to impress, puffs his chest, smiles). You know how it is, get in, give it a go, have a good time.

Davo: I know. Saw you at the boat races . . . remember. Me and the boys chanted you on a bit. Thought you were doing OK. Do ya reckon you'll give the funnels a go tonight?

Paul: Oh shit, I've already had a gut full. (Looks concerned. Doesn't want to be judged as a wimp. Holds his glass up as if to honor someone and downs the remainder of his beer.)

Davo: (laughs) You idiot. You'll be spewing again soon. (Raises his glass to Paul, smiles and downs his glass full.)

Paul: (Nods with approval and pats his gut.) Got to be part of it. What's the point otherwise. *People hold you in high regard . . . the social* *side of it . . . it's the key to the door . . . go out and party hard.* (Looks at Davo waiting for some sign of approval.)

Davo: (laughs again) You'll do alright. (Shakes his head and motions towards the stage. Both move to the side of the stage.)

M.C.: **(Voice over):** Come on lads. Time for a little activity. Get you in the mood for the GIRLS (whistles, cheers, wolf calls). Need a bit of spine, well we've got just the thing for you. (laughter, cheers). Time for the . . . Funnelllll (laughter, cheers, chanting). Come on down . . . MOOSE . . . MOOSE, MOOSE, MOOSE. (Chanting)
(Voice over): Scull, scull, scull, scull.

Chant: Here's to brother MOOSE
Brother Moose
Brother Moose
Brother Moose
Here's to Brother Moose
Who's with us tonight
He's happy
He's jolly
He'll drink it by golly
Here's to Brother Moose
Who's with us tonight
Drink mother fucker, drink mother fucker, drink mother fucker, drink.

Paul & Davo: Chanting, punching fists in the air, jostling each other.

Davo: (turns to Paul) JAFFYs . . . I remember first year. Just like this little prick. I wanted to be part of it, *I wanted to be initiated . . . a lot of people wanted to be initiated into the group.* Look at him (shakes his head and is momentarily distracted by something occurring on the stage).

Paul: Oh gross. He just puked . . . in the bucket. What a loser.

Davo: (looks back at Paul, smiles)
(Voice over): Cocky bastard. I'll fix you!

Davo: Here's your chance, get up there and get noticed. The boys'll be taking note of who's a poof and who's got balls. You don't want to be like the ladies over there, do ya? (Motioning toward the non-drinking observers sitting around the tables in the background.) Go on . . . have a dip . . . that *would get a few bonus points*, get you in.

Paul: What do you mean?

Davo: *It would be something a bit like "Grease" wouldn't it.* You know, *with "the boys" and "the girls" at the top.* (Emphasis on the boys and girls.)

Narrator: Getting "In." Important in terms of gaining acceptance within the "right" circles. "In" groups are in a league of their own and view themselves as being on top of the social ladder. There are certain physical attributes or criteria which, for some, gain them automatic membership into the group. These students have the right image and fit the phys. ed. stereotype of being fit, athletic, "all-rounder type people" who display certain degrees of arrogance and confidence. If you don't gain membership based on any of the physical criteria then there are other factors which can be used as part of a trade process in order to help you "develop" the right image and fit into the "In" group identity. To this end, consideration for membership may be based on being able to use humor and dropping a few one-liners, socializing regularly and impressing "in" group members with being stupid, drinking lots or even attaining "legend" status. One student said that, *if you vomit . . . people sort of put their hands up in the air and say . . . Go for it! . . . That's a legend and helps you be accepted more by the group.* One of the most important factors for survival and on-going membership in the boy's group is to conform to their norms, make sure you have the right attitude and remember to go along with the "anything goes" mentality because acceptance means that, *you are a strong individual within a strong group.*

Paul: (laughing, looking around for someone else) What about him? (points up to stage).

Davo: (Turns back to the stage and cheers out to the next participant) GO BARNEY!!! Now he's a ledge! Believe me, *if you got a reputation you gotta uphold it* and this guy certainly does.

Paul: **(Voice over):** Don't be a pus! This guy is a legend. He obviously thinks I'm OK. Why else would he be talking to a JAFFY?

Paul: OK, I'm in. I'm next. You watch me.

Davo: (calls out to the officials on the stage) He's next boys . . . See how many funnels he can take. . . . And ah, give him the full treatment. (Davo smiles, raises his fist, punches the air).

Davo: **(Voice over):** YES . . . let's see how the little prick goes.

M.C.: **(Voice over):** And here we have?

Paul: Paul.

M.C.: **(Voice over):** Well Paul. Out for a bit of PUSSSSEEE. (Chanting, hooting.) Time to drink.

Davo: (stretching to see, calls out to his mate on the stage) HE'S done it already! LOOK AT HIS GUT!

Davo: He looks like Homer Simpson. . . . Look at him go! Paul Simpson . . . Simpson . . . (smiles) yeah, HOMIE.

Davo: Hey guys . . . HOMIE. HOMIE, HOMIE, HOMIE. . . . (chant changes from drinking song, to HOMIE. Crowd roars) YEAH, (jumps up and punches his fist into the air. Paul stumbles down from the stage and falls at Davo's feet).

Paul: (looks up at Davo who starts to bend down to him) I didn't even puke.

Davo: (slaps Paul on the back, jostles him) mate, you're a legend (looks up and raises his fist . . . laughs) LEGEND. (Bends back over and helps Paul to his feet.)

Paul: (groggy) I didn't puke. Where's me mates, Scottie. . . . Where's . . . . (Spots his mates over toward the back of the room, waves and clenches his fist high in the air.)

Paul: Yes, they saw me. (Looks back at Davo.)

Paul: How's that you bastard . . . . I didn't even puke.

Davo: (gently punching him in the gut, laughing) You soon will son, you soon will. (Steps back from Paul, laughs.) You like the piss don't you. A real little piss drinker.

Paul: (bent forward, holding his stomach, looking towards Davo.) Yep, I sure do.

Davo: You idiot. Like a little more? (pointing to his crotch, laughing.) Gunna puke yet.

Paul: What do you mean?

Davo: Piss, *there was beer and a bit of piss . . . in a few of the funnels.* You took it all. Initiation old son. You're a ledge.

Narrator: "In" group members and more senior members of the boys' group hold the responsibility for initiation activities. Initiation is done in order to identify those students who can party hard, act up and be generally accepted into the wider phys. ed. student social group, but also, selection into the "in-boys" group. Initiation activities often revolve around drinking games, designed for a mild degree of embarrassment and to make sure that the JAFFYs are inducted into understanding the "way it is" in the phys. ed. student social world.

Davo: Just you wait, there's more to come. Gotta pick up. You know how it is, *all the girls are drunk, vulnerable and the boys are, well,* just look around (looks around the room), *drunk, ready. You know . . . WANTING!!* (wiggles his pelvis).

Paul: Oh shit!! (Grabs his gut and falls forward onto his knees, breathes deeply, trying not to be ill.)

Paul: **(Voice over):** Can't be sick . . . can't . . . gotta take it . . . (looking up) Oh shit, what about Kate? Pick-up! . . . . Kate'll . . . Oh, my gut!

Davo: Come on mate . . . . *It's an expectation. Like it's a beer and a smoke . . . It's a beer and a girl.* Gotta get in there. Come on . . . you *know we're just walking hormones.*

Davo: (steps back to Paul, puts his hand on his back) You right mate? (Looks up, strains to see into the crowd. Jumps to his feet.) HERE COME THE GIRLS!! (raises fist into the air and starts to chant) PUSSY, PUSSY, PUSSY, PUSSY.

Paul: (looks up at Davo, gets up and starts to chant) PUSSY, PUSSY, PUSSY.

The ethnodrama produced by Brown (1998) highlights how positions, rules of membership, rites, and rituals are developed in relation to maintaining hegemonic masculinity in PETE. As Brown notes, "Embedded within these hegemonic practices is the subordination and silencing of alternative voices and masculinities: a silencing which is acutely obvious in *Boys' Training*" (p. 84). How humor plays an active role in this subordination is also highlighted. But other interpretations are possible, and this invitation to multiplicity becomes even greater when any play is performed before an audience. Unfortunately, Brown does not tell us if, how, or where the play was performed or what the reaction of the audience was. It would be wonderful to perform this play in front of jock and nonjock audiences, as the pedagogical and analytic potential of this drama is enormous.

Problems of Moving From the Text to the Stage

According to Denzin (1997), in the moment of performance, ethnodramas have the potential of "overcoming the biases of ocular, visual epistemology. They can undo the voyeuristic, gazing eye of the ethnographer, bringing audiences and performers into a jointly felt and shared field experience" (p.94).

The movement from written text to an actual staged performance is complex, difficult, and time consuming. First, as I have indicated, comes the research process itself, which involves the collection of data and an appropriate form of analysis. Next comes the transformation of this data and analysis into a dramatic script of some kind. All this is difficult enough in itself, but perhaps possible for some qualitative researchers who have well-developed writing skills. Moving into the realm of performance, however, takes the researcher into different territory that includes casting, directing, performing, and staging (see McCall, 2000; Wellin, 1996). These are not skills that researchers are ordinarily equipped with. Reflecting on his own involvement in *Handle With Care?*, Gray (2000a) notes how when he took his first steps toward developing a dramatic production he lacked even the basic tools for this kind of work and was blissfully unaware of the challenges that awaited:

> I have thought long and hard about qualitative research issues, and am conversant with the major controversies. But I am not familiar with the theatre world, although I have been on a crash course this last year. In my naivete I thought I could write a script for our production, without too much of a stretch. After all, I have published lots of articles in academic journals, and have a longstanding interest in the creative arts. But it became quickly apparent that I knew nothing about staging and was oblivious to the nuances of visual presentation. For too many years I had been held captive by written text.
>
> *Gray, 2000b, pp. 380-381*

As previously mentioned, Gray and his research team formed a partnership with a theatre program for older adults at one of the universities in the city. The artistic director of this studio agreed to provide artistic leadership in writing, developing, and staging the dramatic production. According to Gray, Ivonoffski, and Sinding (2002), such collaboration is rarely sought. More typically, a social scientist writes a script based on his or her research and then either performs it him/herself or looks for other people to perform it. Gray, Ivonoffski, and Sinding emphasize that while the model of a single researcher/writer/director may be warranted for that rare individual who is multitalented, "in other cases it likely leads to bad theater. And by bad theater we mean representations that fail to deliver the promise of an engaging and visceral connection to the research material" (p. 62). Consequently, qualitative researchers would be well advised to seek collaboration with experts in the expressive arts to transform their findings into a different medium. Such collaboration would not only increase the likelihood of producing an effective drama, but it would also show due respect for others' artistic skills. As Gray points out, this collaboration needs to be handled sensitively.

> My experience may speak more broadly to researchers newly exploring the dramatic sphere. It is worth consulting with experts in the expressive arts. It is critical that we acknowledge not just our own lack of training, but the skill sets others have spent years crafting. If the arts are to avoid becoming a site of imperialistic expansion by researchers, we must negotiate our entries with care.
>
> *Gray, 2000a, p. 381*

Reflections

According to Gray, Ivonoffski, and Sinding (2002), virtually any qualitatively orientated study focusing on participants' own telling of their life situations can provide a foundation for later dramatic work—although the parameters of the study may limit what the drama can subsequently address. Producing a drama does not exclude the production of more conventional articles about a given topic. For example, with regard to their own research on metastatic breast cancer, Gray, Greenberg, et al. (1998) first undertook a standard thematic analysis, wrote it up as a realist tale, and got it published. The drama *Handle With Care?* came after this.

As I have tried to indicate in this chapter, ethnographic drama holds enormous potential for qualitative researchers in terms of disseminating their findings to diverse audiences and effecting social and individual change. It should, therefore, be appealing to critical researchers with emancipatory interests. Furthermore, ethnographic drama enables researchers to understand and analyze their data in a different way—particularly if the researcher should decide to act in the drama. This shift would contrast with the role usually constructed by and for social scientists, which—despite recent moves towards participatory research, the cocreation of interviews, and the inclusion of the researchers' subjectivity—remains that of the observer.

Qualitative researchers often enter carefully into the world of participants, giving much attention to boundary issues, ethics, power, relationships, and so on. Normally, researchers make their exit to analyze, interpret, and write. Some return to the participants to engage in various forms of respondent validation to see if they "got it right." More critical researchers may choose to stay longer and return more often to work directly with groups to bring about change. But this is rare. Most often, qualitative researchers end up operating at arm's length from much of what they study. For Gray (2000a), this is nothing like standing in the participant's body, playing his role, speaking her words, putting oneself on the line. For Gray, becoming a performer in *Handle With Care?* meant he could not remain at arm's length. "I am into it with my whole body. And it connects me to the lived reality of my research participants in new and richer ways. I feel the desperation of those husbands and sons in my bones" (p. 382).

This more embodied way of knowing for both the researcher and the audience would seem particularly attractive to fields like sport and physical activity where the body is taken to be central but for the most part has been curiously missing, absent, or invisible (especially disabled bodies). As I have argued elsewhere (Sparkes, 1997b, 1999a), much recent theorizing about the body has tended to be a cerebral, esoteric, and ultimately disembodied activity that has distanced us from the everyday embodied experiences of ordinary people. Where bodies have been focused on, they have been heavily theorized bodies, detached, distant, and for the most part lacking intimate connection to the lived experiences of the corporeal beings who are the objects of analytical scrutiny. This is not to deny that in certain contexts this kind of work has a part to play in advancing certain ways of understanding the body in culture, sport, and physical activity. The dominance of such studies in the literature is symptomatic, however, of what Wacquant describes as "the discursivist and theoreticist bias of the recent sociology of the body" (1995, p. 89). For him, this bias tends to ignore actual living bodies of flesh and blood. In essence, what Leder (1990) has called the "lived body," as opposed to theories about it, remains curiously, conspicuously absent from contemporary sociological and psychological literature.

Against this backdrop I have raised the question of how we might begin to take more seriously the lived body, the phenomenological and subjective experiences of those involved in sport and physical activity. I have also asked how we might begin to understand the multiple and diverse ways in which people experience their bodies and how these interact to shape identities and selves over time and in specific contexts. In response to these questions, I have suggested that one way forward is to focus on the body narratives or stories told by people. Now, I would like to add another step—that of transforming these body narratives and stories into ethnographic drama or research-based theatre as a way to extend our understanding of bodies beyond abstract theorizing, by including not just the bodies of our research participants, but also our own bodies as researchers-performers-audience.

Accordingly, researchers focusing on the impact on individuals of career-ending injury or illness in sport, for example, or on the acquisition of a disability through sport, can analyze their data and disseminate their findings via the conventional realist tales or even call on autoethnography or poetic representations. They might also, however, consider transforming their data, when appropriate, into ethnodramas. If done well, this transformation could result in the findings reaching a much wider audience than has been the case so far. This more diverse audience would include health care professionals, coaches, teachers, family members, educational administrators and policy makers, support groups, and last but not least the study participants themselves and others like them who have had similar experiences. If done well, a dramatic production of the data could raise issues and generate insights that would lead to both individual and institutional transformations.

Of course, as Gray, Ivonoffski, and Sinding (2002) emphasize, the creation of a drama in itself in no way ensures that social science research will be more broadly accessed or that researchers will make a difference in the world. That is, if the ethnodrama is done badly, then none of its potentials will be realized. In fact, it might turn people away from engaging with the important issues raised by the data. Therefore, given the lack of knowledge and skill that most researchers in sport and physical activity are likely to have in drama or theatre, it would seem appropriate to seek guidance, support, and collaboration with those who do have specific expertise in the performing arts as part of this venture.

The need to work collaboratively with those who have expertise in the performing arts becomes essential when the risks of producing an ethnodrama are considered. For example, acting in an ethnodrama can evoke powerful emotions that can be disturbing for those involved. Furthermore, ethnodramas can inadvertently place vulnerable audiences at risk (see Mienczakowski, 2001).

The move toward a performance medium requires that many researchers extend themselves well beyond the roles they feel secure with. As a consequence, further risks are involved. For example, the time taken in producing high-quality ethnodrama is time taken away from standard forms of production, such as the written article for refereed journals that is the gold standard of academic credibility and career advancement. For some, and here I am thinking particularly of young academics entering the field, the institutional pressures to "perform" to the standard script of university life might prevent them from experimenting with ethnographic drama. They will certainly need support and encouragement from senior colleagues to proceed. Likewise, if ethnographic drama is to be viable, then universities need to find ways to recognize and reward efforts to increase the relevance, and not just the number, of such projects, as well as ways to disseminate findings beyond academic audiences.

Furthermore, as Gray, Ivonoffski, and Sinding (2002) point out, money is a problem for those interested in research-based theatre. They suggest that it makes no sense to discuss the merits of this approach without acknowledging the formidable financial barriers. Indeed, the funding for the original studies on which their dramas were based came from major research funding agencies. As Gray, Ivonoffski, and Sinding recognize, however, these same agencies are unlikely to approve funding to mount and travel a theatre production or to pay for a professional theatre director as part of research plans. Indeed, Gray, Ivonoffski, and Sinding had to turn to cancer organizations and corporate sponsorship to fund this venture.

In summary, the transformation of qualitative data into ethnodrama has much to offer, but there are also risks that researchers need to be aware of as they move toward this genre. My hope is that for some scholars the benefits and rewards will outweigh the risks and that greater use will be made of ethnodrama in sport and physical activity in the future.

Chapter 8

Fictional
Representations

The tales considered in the previous chapters of this book call on various literary and rhetorical strategies as well as diverse fictional techniques. For example, realist tales rearrange time when quotes from interview transcripts are presented out of sequence. Likewise, fictional forms of representation are often used to protect the participants in the study. Coffey and Atkinson (1996) point to the common practice in anthropology, when publishing accounts, of preserving confidentiality by disguising the collective and individual identities of the respondents. In addition to the normal conventions of ethical reporting, such as the use of pseudonyms, nonessential details are often deliberately falsified. Furthermore, because people's research topics can be instantly recognizable to fellow audiences, as can their individual careers in many cases, it often becomes necessary to change details of the research, such as its subject matter and geographical location; to alter details of specific departments; and to fudge who said what. Often, several actors are combined into one, or one is split into several. According

to Coffey and Atkinson, "These little subterfuges are performed primarily for ethical reasons" (p. 128).

In recent years, however, there have been calls for researchers to become even more bold in their use of fiction and to make it more central in their writing. For example, L. Richardson and Lockridge (1991) argue that lived experience is known through the five senses, and the good fiction writer creates a world through the five senses. Therefore, if we want to render lived experience, we have to learn fictional techniques.

With regard to rendering lived experience, others have become increasingly dissatisfied with the limitations on understanding and communication imposed by more traditional forms of reporting; they have both advocated and used what can loosely be described as *ethnographic fiction* and *creative fiction* to represent their research findings in the form of short stories (e.g., Angrosino, 1998; Banks & Banks, 1998; Banks, 2000; Ceglowski, 1997, 2000; Clough, 1999; Diversi, 1998; Dunbar, 1999, 2001; Ellis, 1993, 1995b, 1995c, 2001; Finley & Finley, 1999; K. Frank, 2000; Krizek, 1998; L. Richardson, 1997; L. Richardson & Lockridge, 1998; Tierney, 1993). Of these, several have been explicit about their reasons for choosing fictional forms of representation.

Why Choose Fictional Representations?

As one of his justifications for choosing to use a fictional genre, Angrosino spoke of the ethical concerns inherent in conducting ethnographic work, especially when it involves vulnerable groups:

> The ethnographer is always in danger of betraying trusts, and no amount of informed consent preparation can entirely undo the damage suffered when an "informant" claims that he or she didn't really mean to say this or that, or that he or she didn't expect it to look so shocking when in print. If even "normal" informants aren't really prepared by informed consent, how much less will those with mental disabilities be able to assess the consequences of their consent? The more I can do to disguise individual identities by creating composite characters, shifting scenes, and the like, the more comfortable I can feel that I have gotten to the heart of the situation without unduly breaking any confidences.
>
> *Angrosino, 1998, p. 101*

Besides ethical concerns, there are other reasons for turning to fictional representations. For example, reflecting on her use of fictional techniques to represent the findings of her fieldwork in contemporary strip clubs in the United States, K. Frank comments, "When factual representation obscures possible alternative interpretations, the explicit use of fiction might be appropriate and evocative" (2000, p. 482). For her, "There is a possibility of portraying a complexity of lived experience in fiction that might not always come across in a theoretical explication" (p. 483).

A similar concern led Krizek (1998) to convey the findings of his research on the gendered socialization of sports memorabilia fans in fictional form. He chose this format because it allowed his readers to meet his cultural participants (the people of his "ethnos") in the voice, emotional textures, and multilayered immediacy of the participants' own experiences, in a way not possible through either scientific writing or favored alternatives such as realist or confessional tales. As such, fictional techniques provided him with a means to condense experience in a way that altered time and spanned place. Similarly, during four years of fieldwork with street kids in Brazil, Diversi pondered how he was going to be able to represent them without losing their voices, the three-dimensionality of their humanness, and the mystery that surrounds their lived experiences.

> Early on, I decided that I did not want to represent these street kids' experiences from a theoretical perspective, for that would inevitably bury their voices beneath layers of analysis. . . . I decided to transgress the boundaries of traditional forms of writing in the social sciences . . . by representing the kids I met in my fieldwork through the short story genre. Based on my field notes and on reconstructions of lived experiences I shared with the kids, I employed short story techniques such as alternative points of view, dialogue, unfolding action, and flashback to attempt to create the tension, suspense, delay, and voice that compose a good short story and that are inseparable from lived experience.
>
> *Diversi, 1998, p. 132*

Representing research as fiction, according to K. Frank, "can also affect readers at an immediate and emotional level" (2000, p. 483). Likewise, Rinehart (1998a) suggests that through fiction the reader may *viscerally* inhabit a world, so that ethnographic fiction can both instruct readers and engage their feelings. On this issue, Diversi notes,

> The short story genre has the potential to render lived experience with more verisimilitude than does the traditional realist text, for it enables the reader to feel that interpretation is never finished or complete. . . . [Short stories] have a unique potential to bring lived experiences unknown to the reader closer to his or her own struggles for humanization. . . . Dialogues and descriptions (of places, smells, looks), which are integral parts of short stories, have the power to

move readers from abstract, sterile notions to the lively imagery of otherwise distant social realities.

Diversi, 1998, pp. 132-133

Banks and Banks also argue that the emotional texture of experience is best captured via fiction: "What fiction can do that no other sort of expression does is evoke the emotion of felt experience and portray the values, pathos, grandeur, and spirituality of the human condition" (1998, p. 17). In a similar vein Banks argues, "Emotional ties and individual subjectivity are precisely the sorts of understanding that fiction can help us to share" (2000, p. 400). Indeed, Angrosino states that his main reason for using fictional literary techniques was to "replace the stylistic appurtenances of 'objective,' scientific ethnographic discourse with a style that reinforces the shared subjectivity of experience" (1998, p. 265). He argues that fictional narrative "has been favored by those interested in 'meaning.' And the full deployment of the techniques of fiction . . . is a logical way of achieving a representation of the quest for meaning" (p. 102).

K. Frank (2000) points out that, in addition to portraying the complexity of a situation and affecting readers at an immediate and emotional level, fiction can potentially be used to broaden the researcher's audiences. She suggests that because fiction lacks specialized jargon and is able to involve the reader in a sensory manner, it can be more enjoyable to read than academic writing. Frank notes how she began to publish some of her fiction in local adult entertainment trade magazines and quickly realized that people were reading her stories who would never have picked up an academic journal.

K. Frank (2000) acknowledges how representing research findings in fictional form can provide another valuable *analytical* dimension to the project. Frank turns to fiction—precisely because it is not the same as the other forms of academic writing that she does—to work out problems for which she is unable to find the appropriate theoretical language or framework. This might be because she is not yet quite sure where to locate her ideas in the existing literatures, due to being too "close" to the experience that she is trying to think through, or because she feels that her attempt to translate a certain experience into academic language is incomplete or deficient in some way. Accordingly, she comments, "In attempting to render a character reliable and believable, I can also learn a great deal about the systems and environment in which this fictive person's actions become meaningful as well as about my own questions, assumptions, and emotions" (p. 486). Thus, ethnographic fiction can allow researchers to think about the same data in new ways, and it has the potential to restructure their ideas about the issues they study.

A final reason for choosing to experiment with fiction is simply the *desire* to do so. As Banks (2000) points out, writers have affections for genres of writing and are drawn to produce in their work texts that affirm those affections. Drawing on the thoughts of Krieger (1991), Banks suggests that rather than asserting that we choose to write in a specific genre because of its superior power to explain the world, it might be more honest just to admit that we are fond of the genre. Banks admits to choosing fiction because he desires the challenges, freedom, immediacy, and palpability of creative writing; because he is attracted to the charm of a single case that evokes the universal; and because he enjoys affiliating with the tradition that values truth over facticity. As he states, "I sometimes also choose fiction because I am not compelled to choose it; the rigidities of the tradition's prescribed genres of research writing have always felt confining to me, as dictatorial and self-protective. Risking the choice to write fiction seems like choosing emancipation" (p. 404).

Similarly, K. Frank comments, "In writing short fiction, I have a certain freedom: I do not have to make any conclusions (yet), I do not need to neatly tie up loose ends or anticipate critiques, and I do not need to claim any authority for myself besides that of a storyteller" (2000, p. 486). Thus, plain desire and a sense of freedom would seem reasons as good as any other for choosing ethnographic fiction.

Being There

According to Denison and Rinehart, ethnographic fictions are works "grounded in everyday, concrete, and specific events and research protocols, utilizing fictional strategies to make their conclusions more explicit" (2000, p. 3). As such, it is interesting to note how often the producers of ethnographic fictions are keen to emphasize that their fictionalized representations are based on the author either "being there" in the action as a participant observer or having generated data by other methods in a systematic manner. For example, K. Frank (2000) in her research on the sex industry makes clear her role as a participant observer who sought employment as an entertainer in five different strip clubs that occupied different positions on a status hierarchy. She also points out that she conducted "30 multiple, in-depth interviews with the regular male customers of the clubs as well as . . . shorter interviews with nonregular customers, dancers, managers, advertisers, and other club employees" (p. 481).

Justifying his choice of genre, Diversi (1998) makes it clear that his short story is based on four years of fieldwork—that is, about events that actually happened. Likewise, Krizek is quick to point out the following:

All individual details, characterizations, and experiences recounted in my narrative were witnessed by me at some time during my data collection, although the events did not transpire in the precise order in which they are written. The dialogue is a composite of various interactions written to represent the categories, themes, and cultural understandings uncovered in my "doing" the ethnography. As such the narrative is fiction if fiction implies that the incidents did not unfold as specifically told. The writing is nonfiction if nonfiction signifies that the events and conversations written are compressed from a number of "real" interactions either witnessed or experienced. Be it technically fiction or not, I understand these people and their cultural world . . . and am writing to share that understanding.

Krizek, 1998, p. 94

In a similar fashion, even though he seeks to encompass the truth of various crises in schooling via fiction, Clough (1999) notes that the story he constructs is derived in some measure from "actually" recorded and transcribed interview data along with his own less systematically recorded experiences. Finley and Finley (1999) also point out that the stories they present of the experiences of homeless youth are based on audio-taped conversations.

In making claims to have "been there" and in gathering data in a systematic fashion before producing fictionalized accounts, these scholars display a trend identified by L. Richardson (2000). Here, social scientists who claim their work is fiction usually encase their stories—whether about themselves or groups or cultures—*in settings they have studied ethnographically*. In so doing, these authors appeal to a particular kind of authority, truth, and trust that goes with their being present as witnesses, as opposed to making up their stories purely from imagination.

It would appear that claims to having "been there" are an important, defining feature of any tale that calls itself an "ethnographic" fiction. Such claims are also important in terms of authenticity and credibility in the face of challenges to stories that are creative in their use of fictional techniques but nonfictional in nature. In this sense, there is a connection to notions of descriptive or interpretive validity in traditional realist tales that, according to Agar, are based "almost exclusively on the author's own 'being there' experience (fieldwork)" (1995, p. 7).

Some authors seem reluctant to relinquish all claims to "factual" sources when calling on fictional forms. That is, mixing "fact" and "fiction" can create its own tensions. For example, Ceglowski (1997) notes that although she employed writing techniques borrowed from fiction, including character development, plots, use of dialogue, and a strong storyline, all the events she described were found in her field notes. As such, Ceglowski was uneasy with the term "fictional representation" and preferred to think of her work as employing fictional techniques, such as alternative points of view, deep characterization, third person voice, and omniscient narration.

Similarly, Barone notes how, in his own efforts at educational criticism, "I have felt a tension between my pledge to refer adequately to qualities located within the research setting and the enticements of novelistic modes of fiction" (2000, p. 151). Essentially, according to Angrosino, the dilemma boils down to this:

> We want readers to experience something of what we have experienced, or to experience something of what the people among whom we conducted our research have experienced. We want to import readers into the immediate apprehension of life as it is lived, rather than life as it is analyzed and dissected though the language of positivistic science. And yet many of us were trained to reassure the readers that "this is what *really* happened."
>
> *Angrosino, 1998, p. 97*

How this tension is resolved depends on how far authors are prepared to push their work and make their authority claims along a continuum that ranges from creative nonfiction to creative fiction.

Creative Nonfiction

In making it clear that their stories are based on real events and people, the authors mentioned previously signal that their work is grounded in the tradition of "literary nonfiction," or "creative nonfiction" (see Agar, 1995; Barone, 2000; Denzin, 1997, 2000; Eisner, 1991; Neumann, 1996; L. Richardson, 1997; Rinehart, 1998a). According to Agar, creative nonfiction is fiction in form but factual in content. As Barone notes:

> For an educational portrait to have real value, it is essential that authors use as their material the actual, particular, specific phenomena confronted in the research setting. . . . Unlike the artist, the evaluator is therefore not entirely free to disregard literal "truth." If such an educational portrait is to be convincing, especially to those familiar with the particular research scene, then not least of its virtues must be accuracy. Its characters and setting must be actual, not virtual. Their descriptions should consist of a host of personality indicators, of physical attributes and characteristics of human behavior, in actual incidents, recorded comments, and so on.
>
> *Barone, 2000, pp. 24-25*

Thus, the reader of creative nonfiction needs to know, or be able to presume, that the events actually happened but that the factual evidence is being shaped and dramatized using fiction techniques to provide a forceful, coherent rendering of events that appeals to aesthetic criteria (among others) rather than simply being reported. Writers of creative nonfiction, Barone (2000) notes, use vivid description and other literary techniques to allow the reader to participate vicariously in the events described. Their language is often evocative and metaphor laden and, instead of adopting the traditional role of the objective, factual, detached reporter who "thinly" describes events, "these writers seek to penetrate the personalities of real characters, unveil aspects of their experiences, explicate the social meaning underlying important events in an artful, powerful manner that is at once literary and artistic" (p. 28).

Moreover, writers of creative nonfiction not only describe events vividly but also interpret and present them from a particular point of view. That is, they make judgments and act as critics. As Diversi comments, "To be sure, I am the author of these stories, and, as such, have made important choices in the writing process that both carry my own interpretation of lived experiences and define the possibilities of the reader's interpretations" (1998, p. 133).

In summarizing the fictional techniques most commonly used in creative nonfiction, Agar (1995) points to the *scenic method*. Here, the writer *shows* rather than *tells*. According to Agar, "Situations are recreated for the reader, so that he or she can see and hear, smell and touch, listen to the dialogue, feel the emotional tone. Detailed scenes pull the reader in, involve him or her in the immediacy of the experience" (p. 118). There is the added attraction that the scenes are "real" and not "imagined."

Another technique is that of *character development:* "The writer centers the story on a few 'rounded' characters, real and complicated characters that the reader gets to know and watches develop . . . from scene to scene" (Agar, 1995, p. 118). The reader may be able to slip into and out of different points of view. Internal monologue expresses characters' points of view, and the writer may also become a character in the story. Creative nonfiction also has to have a *plot*. This is developed as the writer selects and arranges details to build dramatic tension. "Parallel narratives, foreshadowing, and flashbacks are among the devices used" (p. 118). Finally, there is the technique of *authorial presence*. As Agar points out, the creative nonfiction writer (like a novelist) is the voice behind—or perhaps in—the story. In this sense the writer is the organizing consciousness, "the force that makes coherent meaning through skilful rendering of the details, a coherent meaning that, like good literature, should offer moral advice on one of the eternal human dilemmas" (p. 118). As Agar explains,

To find "real" detailed scenes and in-depth characterizations or "real" people, the CN [creative nonfiction] writer does "saturation" or "immersion" reporting. They "live with their characters" to acquire the "true" details out of which they will produce their art. CN writers invest substantial time in research; in a forward or note they will tell the reader the general amount and type of research they have done.

Agar, 1995, p. 119

Van Maanen (1988) notes that *literary tales* are spiked by conventions similar to those just mentioned. He emphasizes that even though writers of such tales are careful to stick to their stories and not tell the reader what to think of the presented material, claims to authority remain based on the writer "being there." "Being there" is also important for the *impressionist tales* he describes. These striking stories are permeated with a self-conscious deployment of more "literary" resources that attempt to startle the audience while drawing on fieldwork experience.

As with other forms of tale, impressionist tales draw on a range of conventions, many linked to the techniques used in creative nonfiction. According to Van Maanen (1988), these include *textual identity*, which uses dramatic recall to draw an audience into an unfamiliar story world and allow it, as far as possible, to see, hear, and feel as the researcher saw, heard, and felt. *Fragmented knowledge* is used so that the tale often reads like a novel. This technique can put off those used to realist tales and jar certain sensibilities for them, because the recalled events have uncertain meanings for readers who are unsure of where they are being taken and why; cultural knowledge is slipped to them in fragmented and disjointed ways. *Dramatic control* moves the author back in time to events that might later give rise to understandings or confusions. In the story world, the fieldworker's readings of these events are important. Hence, the recall is sometimes presented in the present tense to give the tale a "you-are-there" feel.

Creative Fiction

Creative fiction is different from ethnographic fiction. Most important for the purposes of this chapter is the willingness of the former to include things that never happened. That is, to give the narrative imagination free rein and invent people, events, and places with a view to crafting an engaging, informative, and evocative story. These inventions can of course be based on real situations, but there is no necessity for this to be the case. With creative fiction, the author is not obligated to inform the reader whether the story told is based on the imagined or the actual. Furthermore, there is no obligation for the author to have "been there" in the field.

Of course, writers may invest time researching a topic in a conventional manner but choose to report their understandings in a form other than creative nonfiction so as to include events that may, or *may not*, have happened. For example, in *Opportunity House: Ethnographic Stories of Mental Retardation*, Angrosino (1998) informs the reader that his stories are based on fieldwork but include events that did not happen and people that did not actually exist. This book is based on 10 years of fieldwork as both a volunteer and a "participant observer" in a home for retarded adults in Florida, among other places. Angrosino has published his findings using more traditional realist formats to illustrate his analytical points, and he acknowledges that these formats are effective up to a point. He notes, however, that what the reader gets in such articles is mainly *him*: "My insights with just a little window-dressing provided by the original words of the people I interviewed. You still don't get much sense of the people as they go about living their lives" (p. 40).

Traditional scientific prose, according to Angrosino (1998), does not enable ethnographers to give readers a sense of people living their lives. To do this, he suggests, requires the use of explicitly literary techniques to create stories that get at the truth of a situation without being explicitly "factual." Accordingly, in *Opportunity House* he uses 12 fictional short stories told in the voices of various community members, as well as in his own voice, to present an ethnographic picture of life. Furthermore, he believes that by using fictional forms he can still cover all the traditional bases of ethnography; that is, say some things about the setting, how people earn their living, how they structure their family and friendship networks, how they relate to the wider society, and how they seek to make meaning in their lives and the lives of others.

> I think that the techniques of fiction—which allow for the creation of composite characters, the invention of situations in which those characters can act and interact—allow me to get at the truth of my experience with OH clients, as well as the truth about their experiences in the world, without having to reveal specific details that might tie the story to any one identifiable client in any one situation.
>
> *Angrosino, 1998, p. 41*

Angrosino's creation of composite characters and invention of situations that did not actually happen signals an important shift away from creative nonfiction toward what he calls *creative fiction* (CF). Talking about his work with composite characters, he comments:

> If I want to illustrate the problem of dysfunctional family relationships as they are experienced by adults with mental retardation, I

might think of several clients whom I know to have negative, destructive relationships with their families. Then I would think of a situation (such as a Christmas dinner) at which those family dynamics could be illustrated in one clear moment, and then imaginatively reconstruct how such people might behave.

Angrosino, 1998, p. 266

Even where the imaginative reconstructions are elaborations on real events, Angrosino (1998) points out that, obviously, the conversations between people when he was not present and the interior monologues of individual characters had to come from what he had learned about people and their attitudes rather than from direct interviews or observational data. Accordingly, Angrosino aspires to create a fictional world as close as he can to the world he worked in for more than a decade. Any appeal to "reality," however, is limited to the sense that things *like these* happened to people *like these*, "not in the sense of this being a documentary with a few literary touches around the edges" (p. 101). He adds,

I hope that fiction can help me convey the truth as I experienced it: that people with mental retardation are a diverse lot with a full range of attitudes, values, and outlooks on life. I can *tell* you that, and it hangs in the air or sits on the page like so much "data." How much better to *show* you the diversity of how people react to situations drawn from, but that are not exactly photographic representations of, "real life."

Angrosino, 1998, p. 101

This move towards creative fiction is supported by Banks and Banks, who, in advocating the use of fictional techniques in social research, comment, "facts don't always tell the truth, or a truth worth worrying about, and the truth in a good story—its resonance with our felt experience, as Walter Fisher says—sometimes must use imaginary facts" (1998, p. 11). Indeed, in comparing creative nonfiction to fiction, Barone points out that artists are "less concerned with reconstructing the literal details of a particular incident, setting, etc.—lest (artists would say) it obscures the 'truth'" (2000, p. 24).

Such sentiments seem to inform the work of Tierney (1993), who, as a researcher subscribing to a critical and postmodern view of the world, became increasingly disenchanted with the static portrayals of organizational life that were being produced in educational research. Accordingly, he created an ethnographic fiction in which, "All the names, characters, and incidents are fictitious, and any resemblance to actual persons or events is entirely coincidental" (p. 314). This fiction was used to explore the background and personalities, along with the conflicts and personal struggles involved, when a university added a sexual orientation clause to its statement of nondiscrimination. Three points were emphasized throughout his story: the conflicting nature of reality, the manner in which change takes place, and the ways organizational change might be interpreted. In making these points through fiction, Tierney notes, "we rearrange facts, events and identities in order to draw the reader into the story in a way that enables deeper understandings of individuals, organizations, or the events themselves" (p. 313). More recently, Tierney (1998) emphasizes that fictional accounts might portray a situation more clearly than more standard forms of representation.

As Banks and Banks (1998) note, with other forms of experimental writing, such as confessional tales and autoethnographies, there is little doubt that the researcher is telling about actual people doing actual things in an actual world. In contrast, they suggest,

> Because fiction necessarily uses the writer's imagination, it vaporizes construct validity and sometimes calls into question reliability in research. It is not necessarily about actual persons, in fact it's not *supposed* to be about actual persons doing actual things in the actual world. Other genres of writing accomplish that just fine.
>
> *Banks & Banks, 1998, p. 17*

Given the nature of creative fiction, any challenges that might be mounted against such stories remain open and different from the challenges that might be mounted against an ethnographic fiction. As K. Frank (2000) makes clear, as a fiction writer, she *invents* events and participants. As a consequence, even though different readers might interpret her stories differently, it is unlikely that one of them will come along and claim, "I was there, and that isn't the whole story." As she points out, this is not necessarily the case when one enters into interactions while conducting fieldwork, or when one creates a fictional story based on data generated in the field. Here, the fictional representation is certainly open to different kinds of challenge from those who participated in the action in terms of, for example, the process known as "member checking." Of course, writers of creative fiction can, and do, check their work by letting trusted friends read their stories to see if they "ring true." That is, they check to see that their story exhibits, for instance, verisimilitude. It remains, however, that producers of ethnographic fiction and creative fiction tend to check their work in different ways.

Ethnographic Fictions in Sport and Physical Activity

In the domain of sport and physical activity, scholars have drawn on the conventions of literary nonfiction, or creative nonfiction, and then produced ethnographic fictions as a means to develop our understanding of key issues. Such works state or infer that the author "was there in the action," that the story is based on "real" people and events, and that "data" have been collected via various means, but that fictional techniques are being used to convey the author's understanding of the situation.

For example, in a recent book edited by Denison and Markula (2002), "Moving Writing": Crafting Movement and Sport Research, there are five chapters that the editors classify as ethnographic fiction. These include Rowe's (2002) highly charged stylization of soccer fanaticism that of necessity locates women on the fringes; Bruce's (2002) story of lesbian issues and sexual identity in sport; Rinehart's (2002) tale of displaced young people; Silvennoinen's (2002) exploration of the ways, as young boys and men grow up in Finland, movement connects to nature and culture; and my own (Sparkes, 2002) fragmentary reflections and memories of my experiences leading up to a third surgical operation on my lumbar spine. In providing an overview of these chapters, the editors describe them as "acts of imagination brought to bear via various realms or settings that have been studied ethnographically" (p. 11; my italics).

Of course, the use of fictional techniques to represent the findings generated by authors who have spent extended time in the field is not unusual. For example, Klein's 1993 book, Little Big Men, is based on a seven-year ethnographic study of four elite gyms on the West coast of the United States. In the service of a predominantly realist telling (see chapter 3, this book), these four gyms are amalgamated (in an analytical sense) to create one fictitious gym, which is the focus of the book.

In contrast, having acknowledged that their knowledge is based on "being there," other scholars are more explicit in their use of stories as a means of conveying their understanding of events. For example, Nilges (1998) critically investigated the lived status of Title IX in one fourth-grade physical education class (n=21) and published her findings as a realist tale in an article titled "I Thought Only Fairy Tales Had Supernatural Power: A Radical Feminist Analysis of Title IX in Physical Education." In this realist tale she comments, "Data were collected over 14 weeks of fieldwork using non-participant observation, field notes, formal and informal interviews, and document analysis" (p. 172). This makes it clear that the tale as told is based on data generated from Nilges' "being there."

In constructing the realist tale, however, Nilges felt dissatisfied with what she was able to "say" and "not say" based on the rules of conventional textualization that structured her original writing. Consequently, in an article titled "The Twice-Told Tale of Alice's Physical Life in Wonderland: Writing Qualitative Research in the 21st Century," Nilges (2001) chose to present her findings in the form of an ethnographic fiction. This choice allowed Nilges to construct an abstract model for understanding the "gendered physical" life of one student, "Alice," by drawing metaphorically from the 1872 book by Lewis Carroll, *Through the Looking Glass and What Alice Found There.*

Alice is the pseudonym given to a popular fourth grade girl who took the lead in her school production of *Alice in Wonderland.* Despite evidence of widespread academic privilege, Alice showed dramatic signs of physical and social alienation in the context of her gender-integrated physical education class. Therefore, to textualize Alice's story and how it intersected with the stories of her 20 classmates, Nilges (2001) condensed her 14 weeks of fieldwork and all the data she collected during that time into the fictive time span of one school day. Thus, the gender journey she takes us on is a fictive construction that temporally sequences "fact" and "fiction." As the tale unfolds, Nilges is able to take Alice and the reader through the looking glass and beneath the surface of "Wonderland" (a pseudonym for Alice's school) to live and critically analyze the limitations of Title IX as a purportedly emancipatory policy.

Others have chosen to bypass the realist tellings and opted early on for ethnographic fictions to present their findings. For example, in an article titled "Pentecostal Aquatics: Sacrifice, Redemption, and Secrecy at Camp," Rinehart (1995) focuses on the issue of ethics in youth sports and, in particular, the ethics between the adult sport-provider and child consumers of sport. Rinehart offers three short stories, each followed by a brief commentary. He comments:

> Mine are both fictionalized and factualized takes, "case study accounts, part fiction and part ethnographic" . . . constructed from my own personal cloth, told in the first and second person, interpreted by me only to be interpreted by you, the "reader" of this text.
>
> *Rinehart, 1995, p. 112*

The same is true with regard to Halas' work, which focuses on the value of physical education and physical activity for troubled youth. In her ethnographic fiction, "Shooting Hoops at the Treatment Center: Sport Stories," Halas provides 24 short vignettes to create a running narrative that, "like a kaleidoscopic lens, focuses in turn on the various lives lived at the school" (2001, p. 79). She too makes it clear that her stories are based on real events and collected data that she reconfigures in the vignettes. In the abstract to her article, she states the following:

> The following series of short stories has been constructed, in part, from the data collected during an interpretive case study of an active living program at an adolescent treatment center/school. Using

excerpts from interview transcripts, fieldwork observations, reflective journal writing, and personal memories (from the author's past experiences as a teacher at the school), this fictional narrative describes how a day at a treatment center/school unfolds, not just on the basketball court, but also in terms of interconnected life issues.

Halas, 2001, p. 77

For seven years, Denison (1994) moved in and out of the world of competitive running. He acknowledges that although his career had many exciting moments, it also had many frustrations when he failed to achieve his goals. All of these made it difficult for him to leave sports behind and move forward in his life. With this personal history as a backdrop, he undertook a study that focused on the retirement experiences of elite athletes. Denison (1996) then chose to present his findings as an ethnographic fiction, rather than a realist tale, in an article called "Sport Narratives." In his notes to this article Denison states:

To complete this study I conducted in-depth interviews, lasting approximately 3 hours, with 12 retired New Zealand athletes (5 men and 7 women). Each athlete had either competed in an Olympic Games or a World Championship in her/his sport. The sports that were represented were: basketball, cricket, cycling, field hockey, netball, squash, swimming, and track and field.

Denison, 1996, p. 360

Thus, Denison (1996) signals to the reader that he "was there" and gathered data from "real" athletes regarding their experiences of retirement. In making such claims, he, like the others mentioned in this chapter, locates his writing in the tradition of creative nonfiction. The end result, for Denison, is three short stories about sport retirement and the problems athletes face in finding some other way to feel good about themselves when their use-by date expires and the fans stop cheering. As an exemplar of this genre, what follows is one of the short stories written by Denison (1996, pp. 352-353).

Spirit

After the 14 finalists for the men's national 5,000 meters championship were spread along the curved white line on the back straight of the Christchurch Athletics Stadium, the starter began to recite his instructions.

"Gentlemen," the old official said in a gruff voice, "take two steps back from the line."

In almost perfect unison, all 14 runners stepped back, among them Steve McDermott. This was Steve's fifth national championship final and he knew the rest of the starter's instructions, the way a criminal knows his rights, so his mind wandered. He remembered how optimistic he was at the beginning of the season, how he was convinced that this was going to be his year. "This is it," he told himself every day, "this is the year I win the nationals." He even wrote his goal on a scrap of paper and folded it up and carried it in his wallet. But lately Steve didn't fancy his chances. For some reason, toward the end of races, when it got tough and began to hurt, he would slow down and start to think about places he'd rather be; drinking with Martin and Andy, seeing a movie with Bridget, anywhere else.

This wasn't Steve's first slump. There was a time back in high school, and again when he was at university—then he couldn't even finish his workouts. But things were different now. He was 26, questions ran through his head: "Am I getting too old? How much longer can I keep running? What else can I do?"

Steve heard the starter's last words—"Good luck, gentlemen"— just before the gun sounded. Quickly he scrambled for position and settled into fifth place, 10 meters off the lead. With 7 of the 12 laps completed, Steve and two other runners broke away from the field. The announcer pronounced it "a definitive move." Steve knew the national champion would come out of this pack. To maintain contact with the two runners in front of him, Steve imagined that a thick blue beam of light connected him to them, a powerful force towing him around and around the red tartan track. With 3 laps remaining, Steve contemplated taking the lead, but he hesitated, believing it might be too early.

As the three leaders swept past the packed grandstand for the penultimate time, the noise from the crowd was deafening. All along the edge of the track people were hollering and clapping their hands. Many of them were screaming Steve's name, urging him on. Behind them the sun was setting below the Canterbury foothills, and Steve glanced up for a moment to admire the burning red and orange horizon. The fiery sky, he thought, was a striking backdrop to the dusty, brown foothills. When Steve looked down again, the blue beam that had connected him with the runners in front of him had disappeared. Suddenly he felt himself drifting back, losing contact. In desperation he began to pump his arms madly, like a drowning man reaching for an imaginary lifeline. Then, with a lap to go, the aroma of grilled sausages and onions cooking over by the concession stand reminded Steve of how hungry he was—he had only had tea and toast for breakfast and lunch. "I'm looking forward to some fish and chips and a few beers tonight," he thought to himself. Then he remembered, "All I gotta do first is finish this damn race."

Creative Fictions in Sport and Physical Activity

Various scholars in the domain of sport and physical activity have begun to experiment with creative fiction. There appear to be two strands in this development at the moment. The first, in a fashion similar to that of Angrosino (1998) and Tierney (1993), is where a creative fiction is presented with some commentary by the author that signals its fictional nature and its purpose. The second is where the creative fiction is presented in the form of a story without any commentary. Examples of the former include the stories told by Duncan (1998) and Sparkes (1997a). These are not based on the systematic collection of data, but they do include events that happened, or that the authors think happened, in the authors' pasts. These remembered events form the basis for stories that weave in events that did not happen.

For example, in her article, "Stories We Tell Ourselves About Ourselves," Duncan (1998) provides three short stories based on her own experiences and two stories written by her students, also based on autobiographical experiences. The stories may be classed as creative nonfictions. In combination, they explore how stigmatized bodies are constructed and investigate the feeling-world that is embedded in this process. In this sense, these short stories could have been included in chapter 4 on autoethnography. Indeed, the boundaries between the two genres, when it comes to using fictional techniques, are blurred and permeable—at times indistinguishable.

Duncan's (1998) work, however, for me, makes an important move toward creative fiction. Duncan makes no claims to having conducted systematic research that generated "real data" on the topic, as the authors of ethnographic fictions are inclined to do. Rather, she relies on memories of being there and makes literary additions that do not require her to have been there, so as to make her analytic points via the story she tells. Drawing on Rorty (1989) and his thinking on the "narrative turn," Duncan argues for the importance of stories in particular, and the imagination in general, as tools for increasing our sensitivity to the pain and humiliation of other, unfamiliar sorts of people. She suggests that genres such as ethnography, the journalist's report, the novel, and the movie might point the moral way for us by their ability to convey the suffering of those whom we might be tempted to dismiss as having nothing in common with us: "Fiction in general offers detailed description that shows us the kind of cruelty we inflict on others, often unthinkingly, although no less brutally. Stories allow us to re-envision ourselves as the marginalized Other, and thereby offer us the possibility of moral behavior" (p. 97).

Discussing one of her stories, Duncan notes, "My story about Steven is an adult re-creation of a childhood memory, undoubtedly full of gaps and inaccuracies" (1998, p. 97). That is, it is a mixture of facts and fictions in unknown measure. She later adds, "For myself, the truth of my memory lies in the force of its moral imperative. It has helped me understand the cruelty of stigmatizing others through their bodies" (p. 97). Thus, the stories have critical intentions.

In a similar fashion, my own article, "Ethnographic Fiction and Representing the Absent Other" (Sparkes, 1997a), was not based on any systematic observation or interviews with specific people. Nor can I make any claims to "being there" other than in the limited sense that the fiction draws on some events that happened in my life, or events I was told about by others, before recasting them into a story with a particular purpose.

For example, I invented "Alexander," who is the main character in the story; he is not real in the sense of being an actual person that I interviewed or observed. Various parts of the story, however, draw on my own experience of being a rugby team member and a physical education student and teacher. For example, when I was at university some members of the rugby club submitted the names of their friends to the Gay Society, and while playing for a first-class rugby team I was invited to go "gay bashing" (I declined). Therefore, as I constructed the story I tried to imagine what I would have felt like if I was gay and these events happened. I tried to imagine what it must feel like to be a gay sportsman and physical education teacher in a homophobic jock culture. Of course, the story is written from a heterosexual perspective, and I can only imagine from this frame of reference; the particular character of Alexander is a fictional self-fashioning that constitutes but one of many possible interpretations. The telling of any story reflects something of the teller, so that Alexander's story also provides an autobiographical statement about me as author.

The context that led to the production of my ethnographic fiction is described in detail elsewhere (see Sparkes, 1997a). I produced it for pedagogical reasons, and I had critical intentions, in an attempt like that of Duncan (1998), to speak for the absent other—in this case, gay, male physical education teachers. These concerns about how to represent the absent other so that their voices could be heard emerged as part of a module on "Equity and equality in sport and physical education" that I gave to undergraduate and postgraduate students, where I dealt with issues relating to sexual identity.

These classes were dominated by white, male, heterosexual jocks and superjocks, with support from their able-bodied female counterparts (Dewar, 1990). All of them, and myself as well, were embedded in a sporting and physical education culture permeated by homophobia and heterosexism. Similarly, the students' views of sport and physical education, and the topics they deemed legitimate in this domain, were (and still are) strongly framed by instrumental or technical forms of rationality that, via the hidden curriculum, represent teaching and learning as technical and unproblematic matters rather than as a political, critical, contested, and intellectual endeavor. Not surprisingly, the students I worked with, like those described in an Australian study by Macdonald and Tinning (1995), displayed a strong preference for specific types of *utilitarian* knowledge. They also tended to hold a biologistic view of a person that rests on notions of the body as a machine that can be known in every detail, manipulated, and subjected to the needs of technology.

Given this context, I was extremely concerned about how I was to operate in the lecture-based setting when dealing with the complexities that surround issues relating to sexual identity, homophobia, and heterosexism. For example, how might I address these issues in a holistic manner that would draw on the multilayered and diverse, but too often repressed, subjectivities of the students? How was I to present material? How could I invite discussion in a nonthreatening and supportive manner? Should I invite discussion at all? What does "discussion" mean? Was it up to me to "give voice" to the lesbian students who I knew were in the audience and with whom I had discussed the proposed content of the lectures, and to the other lesbian and gay students who might be in the audience but wished, or were forced, to remain silent? Indeed, just what does it mean to "give voice"? What kinds of relations does this giving imply? What kinds of power and privilege are implied in the act of giving? How does this act lock me into preexisting cycles of domination and privilege and their ongoing replication? Why should the oppressed speak up, anyway?

Taking my lead from the feminist pedagogy of Dewar (1991), I incorporated video material and journal articles that focused on lesbian and gay issues in sport. I also used journal articles and book chapters that focused on the lived experiences of lesbian physical education teachers. I could find nothing, however, that considered the experiences of gay, male physical education teachers. After much thought, I chose to write an ethnographic fiction to represent these absent others to students and to illustrate how their experiences of homophobia and heterosexism might differ significantly from those of lesbians. When I hand the story out, I do not introduce it but simply ask students to read it. This is the story (Sparkes, 1997a, pp. 28-33).

Moments From Alexander's Story

Alexander is in his late 20s. He is 6 feet tall, weighs about 14 stones. His muscular physique bears testimony to many years of hard training coupled with the blessings of good "mesomorphic" genetics. Alexander excels at a range of sports but has been particularly successful in rugby which he has played at first-class level. As a teacher Alexander is dynamic, articulate and outgoing. He gives a lot to the students he teaches in terms of energy and time. They, in turn, think highly of him. So do his fellow teachers who find him a committed and supportive colleague to work with; someone who gives a great deal to the life of the school and has developed the PE program extensively in terms of, for example, developing health-related exercise modules.

In many ways Alexander has all the characteristics of the "good" male PE teacher. All the characteristics except one that is. Alexander is gay. He has only recently explicitly come out to his mother following the early death of his father. The pain of losing his father, a man he idolized in many ways, also acted as a release and a stimulus for Alex to "name" himself and take a new direction in his own life. What follows are some moments from his life in his own words.

> Growing up thinking you are gay in a strong working class family isn't my idea of fun. My Dad had a tough upbringing and sport, especially football, played a big part in his life. In fact, he almost turned professional but it didn't quite work out. He was a real "man's man," a hard man. When I say that, he was hard but fair. But I think he found it very difficult to show his emotions. I can't ever remember him hugging me a lot as a child, not unless I made the first move. He would still get embarrassed when I gave him a hug in public sometimes when I went home for a visit. I think he was simply like many men of his generation.
>
> Sport was our connection, our bonding. He got me playing football for local teams as soon as I could walk and I was clearly talented. Then, when I was eleven I had the chance to get a scholarship to a direct grant school, but it played rugby. It was a quandary for my Dad for a while but to him, and my Mum, getting a good education and "getting out" of the kinds of jobs they were locked into was all important. So I went and, lo and behold, I took to rugby easily. Again, without being big headed it was clear that I was talented in sport and the teachers recognized this. Mum came to watch when she could but she had a Saturday job. My Dad would come and watch every match and I knew he

got a buzz from it when I played well. I remember when I played my first game of top class rugby and I scored my first try. As I ran back I looked over into the crowd at him and you could feel the pride oozing out of him. He just wanted to shout "That's my son!" Funnily enough, even in those situations he found it difficult to show his emotions. After the game he just said, "You did well out there today Alex, a pity about that missed tackle!" He would always deflect any compliment with humor. But that was my Dad.

I miss my Dad like mad. His death has left a big hole in my life. It's difficult to say how much he meant to me. His death was a turning point for me in many ways. You see, although I love him and I know he loved me there was one thing we could never discuss, and that was sexuality. I could never tell him I was gay. He just couldn't have coped with it. You see, in my Dad's world, and particularly in his world of work, the pub and sport, men were always "real" men. That is, they were straight [heterosexual]. As I grew up I learnt from the jokes he would tell his mates, and later to me when I started playing rugby, that "queers," "faggots" and "poofters" were a source of ridicule, something to be despised.

Other than the jokes we never spoke about gay men directly, that is as a topic in itself, but his views slipped out sometimes. I remember when one of the Fashanu brothers [black, professional soccer players], Justin I think it was, "came out" and said he was gay. I thought my old man was going to have a fit. There it was all over the sports page of the Sunday paper. He just could not believe that a professional footballer could be gay. When he did have to believe it because Fashanu confirmed it on the TV, his reaction was hostile. Something along the lines of "He should be banned. I wouldn't want to have a shower next to a queer. No thank you!" The AIDS issue has also brought out the worst in him. When Freddie Mercury [lead singer with Queen] died he just said, "Good riddance to bad rubbish. That AIDS stuff will sort all those queers out." Only a few years ago I remember taking him and Mum out for a meal and my Dad got it into his head that the waiter was gay. Just from the way he talked, my Dad said he was gay. This made him nervous because, in my Dad's head, gay means AIDS, and AIDS can be passed on by someone touching a plate that you eat off! Honestly, it really got to him. How do you fight that kind of ignorance when it's your own Dad? I tried to put him straight on some of the facts about AIDS but without much success I'm afraid.

Now all this kind of stuff was being reinforced by the sports I was involved in, and especially the rugby. At school the big insult you gave another boy was to call him a "queer" or a "gay boy." It was a standard form of put down in the school rugby team, "What are you, queer or something?" And we would all laugh. I would laugh too but inside I felt very uneasy about it all. As I moved up the ranks in rugby the joking went on but I saw another side to all this, a violent side. We had played a game in London and after we had eaten and what have you, some of the guys in the third team who had traveled with us put out a general announcement that they were going to some bars for a bit of "queer bashing." I remember feeling my blood go cold as I watched five big drunken men set off to physically intimidate, perhaps harm, someone they had never met but were going to have a go at just because they were gay. By the time they got back to the coach to leave they were in a drunken stupor. But they remembered enough to tell us that they had taunted and roughed up a young guy coming out of a gay club. I sat there thinking how they would react if they knew I was gay.

And so on it went. I was trying to sort my life out with all this hatred going on around me. I began to feel I was different during adolescence. I learnt the game of going out with girls but I didn't feel right about it. I guess my emotional connections wanted to be made with men but first, I didn't know how to make the connections and second, I was scared to admit to myself that I was gay. It was so difficult to get a fix on. I could hardly go up to my Dad and say, "Hey Dad, I think I'm queer. Can we have a chat about it?" You must be bloody joking! You couldn't talk to the teachers. I got on well with the PE teacher but he was married with kids, clearly straight, and he had never stopped negative comments or jokes being made about gays when we traveled with the rugby team or in the PE lessons, so nothing there for me. Also, I just didn't know anybody who was gay and who was "out." I had no reference points, nobody to talk things over with. It was a bad period in my life in that sense despite all the success I was getting elsewhere on the sports field.

As "A" levels loomed I was asked by the PE teacher what I planned to do when I left. As usual I didn't have a bloody clue. So he suggested doing a degree in sports science or something like that as it would allow me to continue with my sport and get a qualification as well. As they say, "It seemed like a good idea at the time." So I applied, got the grades and got in. It was good being away from home. I missed my Mum a lot, I missed my Dad. But, again, I knew that me "their son" going to university to get a degree made them as proud as hell. In fact, university was a Godsend. It gave me a bit of space away from my family to get to know myself. It wasn't all hunky dory though. I was still locked into the sports culture, or the jock culture I should say. As usual, I heard nothing positive said about gays and the underlying feelings of hostility clearly were still there. For example, at the Freshers' Bazaar where you can join all the clubs some of the big guys from the rugby club "minced over" and joined the Lesbian and Gay Society but gave names of their friends in the rugby team. A source of great amusement in the bar later. I laughed about it while at the same time wishing I had the guts to go and join the society for real. Having said that, at the bazaar it was the one stand that I kept well clear of. Almost fear by association if you like. I guess I had to be more straight than straight to protect myself, to keep the front going.

A couple of months later I was in a pub and I saw one of the guys who was on the stand of the Lesbian and Gay Society. I scanned the pub to make sure that there was nobody there that I knew and then went over to talk to him. He was really great and made me feel at ease immediately. He was a very good listener. I guess he had to be because I spilled it all out, there and then in that pub. It was the first time I had ever talked to another person about myself being gay. It was a big weight off my shoulders. Mark was also good because he knew the kind of culture I was living in, he knew how jocks viewed lesbian and gay issues. So his advice was sensible. He didn't advocate some dramatic "coming out" to the jocks or the rugby team as the consequences would be too severe. I remember laughing and saying that death would be a simpler alternative! No, he simply said it was about time I began to get to know myself and feel good about who I was. So we worked within the framework I was living in. He introduced me to some gay friends and gave me the addresses of gay support groups and some gay clubs. It was all new to me. I was excited and scared at the same time. Eventually, I met Gavin and it all happened. When it did it just felt so natural, so good, so right for me. I had said hello to myself. I wasn't afraid any more. I knew who I was.

Knowing I was gay and feeling good about my sexuality didn't help me much on a daily basis at university. I never came out to anyone on my sports science course, the PGCE course, or in the rugby team. You see, jock culture is based on the notion of straight sex. It's male dominated, and it's about screwing women who are basically seen as objects for the pleasure of the lads. That may sound a bit harsh, but deep down inside that's how it works. I simply played my part when I had to. Most of the time I managed to detach myself from things more than most, that made them see me as "serious" and a bit too much into academic work. But I also had all the right credentials. I could do all the sports well and I was first team player in rugby, that counts for a lot. So when I went to a disco or a club I would dance the night away and leave with a girl to walk her home. The assumption was then made by the rest of the lads that I would, by definition, sleep with her. I never did and simply walked them home. When I felt that the women expected something more from me I made some lame excuse to get me out of the situation. But the point is the boys assumed I had done it. Like, next day they would say "Have a good night Alex?" with a big grin on their faces. I would simply grin back and say "Yes thanks." I wasn't lying to them. I did have a good night, but not in the way they meant. So I got by. The rugby set up was a bit claustrophobic but I used the space that the travel away to rugby games gave me to visit gay clubs in some of the larger cities where I felt nobody would recognize me. I told the lads that I was going to stay with a friend in the city and made my way back to university on my own the next day.

Splitting my life in two carried on when I began teaching. I went in knowing that I would have to stay in the closet as far as teaching PE was concerned. All that has been confirmed by my experiences in this school. Even more so now that Clause 28 [anti-gay legislation] is around. Anyone who comes out now is dead professionally, they wouldn't last ten minutes before the governors got them out on some technicality or other. I remember the boss [head teacher] last year asked me about one of the candidates for a job in another subject because he had put down some qualifications that meant he could help with games. The boss said, only in passing, that the guy was 40 years old and not married which wasn't a good "sign," and that they had to be careful recruiting these days, especially if it was anything to do with kids in PE, not to get the "wrong kind of person." He meant "gay" of course. Single, male, 40, therefore possibly gay. Result, no interview. So can you imagine the headlines in the Sun if they found out that I was a full-time PE teacher and a full-time gay, "Gay PE teacher watches children in showers." Tell me about it!

Having said all that, because I look the part, and because of the rugby and so on, everyone just assumes I'm straight. Nobody has ever asked me, kids or staff, if I'm gay. My masculinity has never been questioned. I've got all the right "visible" credentials. Quite simply, they don't even conceptualize that a gay male PE teacher could exist. They just can't get their heads around that one. Can you imagine it, PE the bastion of masculinity, the makers of real men, has gays in their midst! So I just pass as straight because the question is never raised about whether I'm gay or not. I just let the assumption ride and never challenge it. For example, I flirt with some of the women teachers and we get on well, we have a great laugh. When there is a staff do I usually invite one of my female friends who I've met through the rugby club or the gym I use. We all have a good time and no questions asked. So in that sense being gay has not been a problem for me. I play the game.

In the last few years I've become more and more dissatisfied about remaining in the closet at school and leading this double life. You see, two years ago I met Andy and we have been together ever since. I'm getting sick of denying his existence and talking at school about going off to see my "girlfriend" at the weekend when I'm going to see him. We want to make a go of it and live together. Now clearly, if he moved up here with me that could cause some problems. It wouldn't take too long perhaps before something got back to the school and I would live in fear of that. But that issue won't arise. Andy has his own business and he has to stay put. That means I have to move. So what with my Dad's death, meeting Andy and all that goes with that I have had to come to some crucial decisions about the direction I'm going to go in from now on. I'm getting close to thirty now. My rugby is becoming less and less important to me and I've got into lots of other activities, gym work, just keeping fit basically and I really enjoy it. If I stay in teaching, particularly PE teaching, I don't think I will ever be able to come out publicly and I'll spend my professional life in the closet living a lie. I just don't want that any more.

I don't want to go and work as an employee in Andy's business as I feel the other staff might feel I have an unfair advantage over them. He's out. They know he is gay. So what I've done is got a job with a large sports marketing company in the same city. That way we can be together but I still maintain my independence in a financial sense, that's important to me. I suppose I could have got a job teaching in the area but what would have been the point, it would not be any different from here really. My Mum's been great about it all. She's met Andy and likes him a lot. I feel so good that she knows about me and accepts things as they are even though I don't think she really understands all the implications of me coming out. I still wonder what would have happened if I'd told my Dad. I guess I'll never know. Perhaps it's best that way. So that's it. I'm going to be out as far as you can in today's society, and I'm going to leave teaching at the end of this term. Scary isn't it.

More Stories

As illustrated previously, the stories told by Duncan (1998) and myself (Sparkes, 1997a) both include some commentary that frames the story being told and shapes its interpretation. In contrast, various scholars in the domain of sport and physical activity have opted to let their creative fictions stand by themselves, with no commentary from the authors, no references to other literature, and no "academic" interpretation. Indeed, no claim is made that the events described actually happened at all. It could all be invented. The reader simply does not know. This kind of work is presented as creative fiction, pure and simple.

In a special edition of the *Sociology of Sport Journal* on the narrative imagination, Denison and Rinehart (2000) as the guest editors point to works as exemplars of creative fictions. Bethanis' (2000) story, "The Shadowboxer," describes an adolescent boy learning what it means to become a man and discovering his place in his father's world. In "Believing," Christensen (2000) creates a poignant and desperate picture of corruption and exploitation in collegiate sport. Rowe's (2000) story, "*Amour Impropre,* or 'Fever Pitch' Sans Reflexivity," uses an extended metaphor (what used to be termed a conceit) to show an extreme example of football fanaticism, which is emblematic of sport misogyny. In a story titled "Gift," Denison (2000) casts doubt on just how wonderful it really is for a young person to have the "gift" for sport. His work also raises questions about parental influence, child preconsciousness, structured unfairness, and the emphasis Western nations place on achievement. Finally, there is the finely layered story by Wood (2000), "Disappearing." According to Denison and Rinehart, this story resonates on many levels: "as a demonstration of the hopelessness of subjugation, as a springboard for discussion of eating- and exercise-disordered behaviors, and, finally, as a forum for empowerment" (p. 3). Although the guest editors make a brief comment on each of the fictions mentioned previously, the authors themselves make no direct comment, and the interpretations of the stories remain open, to be used by readers as they see fit.

With the previous points in mind, here is an exemplar of a creative fiction. It is Wood's "Disappearing" (2000, pp. 100-102).

Disappearing

When he starts in, I don't look anymore, I know what it looks like, what he feels like, tobacco on his teeth. I just lie in the deep sheets and shut my eyes. I make noises that make it go faster, and when he's done he's as far from me as he gets. He could be dead he's so far away.

Lettie says leave then stupid but who would want me. Three hundred pounds anyway but I never check. Skin like tapioca pudding, I wouldn't show anyone. A man.

So we go to the pool at junior high, swimming lessons. First it's blow bubbles and breathe, blow and breathe. Awful, hot nosefuls of chlorine. My eyes stinging red and patches on my skin. I look worse. We'll get caps and goggles and earplugs and body cream Lettie says. It's better.

There are girls there, what bodies. Looking at me and Lettie out the side of their eyes. Gold hair, skin like milk, chlorine or no.

They thought when I first lowered into the pool, that fat one parting the Red Sea. I didn't care. Something happened when I floated. Good said the little instructor. A little redhead in an emerald suit, no stomach, a depression almost, and white wet skin. Good she said you float just great. Now we're getting somewhere. The whistle around her neck blinded my eyes. And the water under the fluorescent lights. I got scared and couldn't float again. The bottom of the pool was scarred, drops of gray shadow rippling. Without the water I would crack open my head, my dry flesh would sound like a splash on the tiles.

At home I ate a cake and a bottle of milk. No wonder you look like that he said. How can you stand yourself. You're no Cary Grant I told him and he laughed and laughed until I threw up.

When this happens I want to throw up again and again until my heart flops out wet and writhing on the kitchen floor. Then he would know I have one and it moves.

So I went back. And floated again. My arms came around and the groan of the water made the tight blondes smirk but I heard Good that's the crawl that's it in fragments from the redhead when I lifted my face. Through the earplugs I heard her skinny voice. She was happy that I was floating and moving too.

Lettie stopped the lessons and read to me things out of magazines. You have to swim a lot to lose weight. You have to stop eating too. Forget cake and ice cream. Doritos are out. I'm not doing it for that I told her but she wouldn't believe me. She couldn't imagine.

Looking down that shaft of water I know I won't fall. The water shimmers and eases up and down, the heft of me doesn't matter, I float anyway.

He says it makes no difference I look the same. But I'm not the same. I can hold myself up in deep water. I can move my arms and feet and the water goes behind me, the wall comes closer. I can look down twelve feet to a cold slab of tile and not be afraid. It makes a difference I tell him. Better believe it mister.

Then this other part happens. Other men interest me. I look at them, real ones, not the ones on TV that's something else entirely. These are real. The one with the white milkweed hair who delivers the mail. The meter man from the light company, heavy thick feet in boots. A smile. Teeth. I drop something out of the cart in the supermarket to see who will pick it up. Sometimes a man. One had yellow short hair and called me ma'am. Young. Thin legs and an accent. One was older. Looked me in the eyes. Heavy, but not like me. My eyes are nice. I color the lids. In the pool it runs off in blue tears. When I come out my face is naked.

The lessons are over, I'm certified. A little certificate signed by the redhead. She says I can swim and I can. I'd do better with her body, thin calves hard as granite.

I get a lane to myself, no one shares. The blondes ignore me now that I don't splash the water, know how to lower myself silently. And when I swim I cut the water cleanly.

For one hour every day I am thin, thin as water, transparent, invisible, steam or smoke.

The redhead is gone, they put her at a different pool and I miss the glare of the whistle dangling between her emerald breasts. Lettie won't come over at all now that she is fatter than me. You're so uppity she says. All this talk about water and who do you think you are.

He says I'm looking all right, so at night it is worse but sometimes now when he starts in I say no. On Sundays the pool is closed I can't say no. I haven't been invisible. Even on days when I don't say no it's all right, he's better.

One night he says it won't last, what about the freezer full of low-cal dinners and that machine in the basement. I'm not doing it for that and he doesn't believe me either. But this time there is another part. There are other men in the water I tell him. Fish he says. Fish in the sea. Good luck.

Ma you've lost says my daughter-in-law, the one who didn't want me in the wedding pictures. One with the whole family, she couldn't help that. I learned how to swim I tell her. You should try it, it might help your ugly disposition.

They closed the pool for two weeks and I went crazy. Repairing the tiles. I went there anyway, drove by in the car. I drank water all day.

Then they opened again and I went every day, sometimes four times until the green paint and new stripes looked familiar as a face. At first the water was heavy as blood but I kept on until it got thinner and thinner, just enough to hold me up. That was when I stopped with the goggles and cap and plugs, things that keep the water out of me.

There was a time I went the day before a holiday and no one was there. It was echoey silence just me and the soundless empty pool and a lifeguard behind the glass. I lowered myself so slow it hurt every muscle but not a blip of water not a ripple not one sound and I was under in that other quiet, so quiet some tears got out, I saw their blue trail swirling.

The redhead is back and nods, she has seen me somewhere. I tell her I took lessons and she still doesn't remember.

This has gone too far he says I'm putting you in hospital. He calls them at the pool and they pay no attention. He doesn't touch me and I smile into my pillow, a secret smile in my own square of the dark.

Oh my God Lettie says what the hell are you doing what the hell do you think you are doing. I'm disappearing I tell her and what can you do about it not a blessed thing.

For a long time in the middle of it people looked at me. Men. And I thought about it. Believe it, I thought. And now they don't look at me again. And it's better.

I'm almost there. Almost water.

The redhead taught me how to dive, how to tuck my head and vanish like a needle into skin, and every time it happens, my feet leaving the board, I think, this will be the time.

The Issue of Purpose

I feel uneasy in presenting the examples in the manner I have and in making a neat distinction between ethnographic fiction and creative fiction. As the other chapters in this book illustrate, numerous genres, including scientific and realist tales, call on fictional and literary techniques to represent findings. That is, all research writing is crafted, constructed, fashioned, invented, and composed (see chapters 1 and 4). In this sense, the ethnographic fiction of Denison (1996) is as much a "making" as the creative fiction of Wood (2000). Therefore the boundaries that I have manufactured for the purposes of discussion in this chapter are fragile, permeable, and, most of all, debatable.

As K. Frank (2000) argues, it is possible to distinguish between different kinds of stories in terms of the processes that lead to their production. For Frank, it is important that authors acknowledge the genres they have chosen so that readers are aware of what is being offered to them. On this issue, Frank makes her position clear:

> It is essential, and possible, to distinguish between stories about events told by informants, stories about events recollected from a fieldworker's own experiences, and stories about events that originated wholly within the writer's imagination but that draw on situated knowledge that the writer possesses. . . . I would find it both unethical and irresponsible to use fiction in an academic text without explicitly stating that fact for the reader and differentiating it from sections in which fictional techniques were used to evoke a mood, conversation, or setting.
>
> *K. Frank, 2000, p. 485*

Authors need to be aware of their responsibilities to readers and of the ethics involved when they produce fictions. Of course, as Barone points out, this relates to the manner of fashioning, or *modes of fiction*, favored by authors, and those modes will vary greatly in accordance with purpose: "A novelist with a narrative purpose will select different modes of fiction from, say, the emic-orientated anthropologist who claims to fashion a 'correct' actor-oriented interpretation" (2000, p. 147). As Geertz comments:

> In the former case, the actors are represented as not having existed and the events as not having happened, while in the latter they are represented as actual, or having been so. This is a difference of no mean importance . . . but the importance does not lie in the fact that [one] story was created while [the other] was only noted. The conditions of their creation, and the point of it (to say nothing of the manner and the quality) differ.
>
> *Geertz, 1974, pp. 15-16*

Thus, the *purpose* behind the choice of ethnographic fiction or creative fiction as opposed to another genre is the important issue. For example, despite the giddiness felt lately by researchers on release from their methodological straitjackets, Barone (1997) warns against the use of fiction just for the sake of using fiction. For him, the use of fiction is warranted and literary license is wise "only when employed in the service of a legitimate research purpose. That purpose is the generation of a conversation about important educational questions" (p. 223). A similar point is made by Coffey and Atkinson (1996) when they argue that ethnographic fiction should be used to "construct and convey *analyses* of social settings and social action that are given particular point or are impossible by other means. . . . one must be clear that such exercises have an analytical purpose" (pp. 128-129).

Thorne (1997) also raises the issue of purpose in her discussion of the social impact of artistic qualitative research products and the accompanying obligations. She suggests that, unlike the poet attempting to capture the smell of a rose, or the painter trying to convey an essential emotion in visual form, "qualitative social and health science researchers are engaged in the business of knowledge production for some purpose, and that purpose has very real social consequences" (p. 129). That is, the stories we tell about people have an impact on how they see themselves, how others see them, and how they act toward each other. For Thorne, "our capacity to render research findings using the persuasive traditions of both art and science makes those findings all the more potent in the social world" (p. 129).

This potency, and the responsibilities that go with it, are particularly evident when the purpose of writing is to explore the lived experiences of silenced, absent, and marginalized others. Here, fictional representations have the potential to make a powerful contribution (Barone, 2000; Duncan, 1998; McLaughlin & Tierney, 1993; Sparkes, 1997a; Tierney, 1993; Tierney & Lincoln, 1997). Accordingly, it is interesting to note the reasons Denison gives for producing his own ethnographic fiction.

> I chose to represent my subjects' experiences through stories because I did not want to write about sports retirement in a way that centered on theory, where my subjects' voices would be buried beneath layers of analysis. . . . With fiction I felt I could write about sports retirement in a more evocative way, employing such devices as flashback, alternative points of view, and dialogue to open up the dimension of mystery that surrounds athletes' retirement experiences. Furthermore, with fiction I could avoid closure, enabling the reader to see that interpretation is never finished. . . . stories show instead of tell; they are less author-centred; they allow the reader to interpret and make meaning, thus recognizing that the text has no universal

or general claim to authority; and, most important, they effectively communicate what has been learned.

Denison, 1996, p. 352

A clear sense of purpose is also evident in the work of Halas (2001) on marginalized youth when she states, "My desire to understand what a nurturing and relevant physical education program looks like for troubled youth shaped my thesis research and currently motivates my involvement with the ethnographic fiction I am presenting here" (p. 79). Halas continues, "My goal in writing these stories was to provide a multi-faceted portrait of how a day at the treatment center/school unfolds, not just on the basketball court, but also in terms of interconnected life issues" (p. 79). Similarly, the stories constructed by Duncan (1998) and myself (Sparkes, 1997a) have definite pedagogical purposes. These authors have made purposeful, informed, and strategic choices regarding their use of specific genres. There are several reasons scholars might usefully engage with ethnographic fiction or creative fiction.

The Potential of Fictional Representations

Various scholars have commented on the potential of fictional representations to help us understand and connect with the world around us in different ways. For example, Diversi (1998) argues that fiction can bring the distant social realities of others closer to the reader. For him, "Well-written short stories, which create tensions and voices that sound real, have the power to allow readers to see themselves in the human dramas being represented, even if the specific circumstances shaping these human dramas are different from the circumstances shaping the readers' own dramas" (p. 133). Indeed, Diversi's most optimistic goal is that the stories he crafts "will connect, at least initially, in an emotional realm, human beings living in extremely different social contexts" (p. 133).

This goal is reflected in Barone's (2000) call for educational storytelling and sharing. Here, the aim of storytellers is not to prompt a single, closed, convergent reading but to persuade readers to contribute answers to the dilemmas they pose. For Barone, stories are able to provide powerful insights into the lived experiences of absent others in ways that can inform, awaken, and disturb readers by illustrating their involvement in social processes of which they may not be consciously aware. Sometimes, individuals may find the consequences of their involvement in oppressive processes unacceptable and seek to change the situation. In such circumstances, the potential for individual and collective restorying is enhanced (Sparkes, 1994a). Nonetheless, the notion of stories and storytelling does not sit easily in many academic circles.

According to Barone (1995), the prevailing vision of what constitutes legitimate scholarship may partially explain the relative scarcity of published examples of fictional and nonfictional stories about school people written entirely in the vernacular. Rather, stories by and about teachers and students in professional journals are nearly always folded in didactic material aimed at exacting scholarly meaning from the accounts. Barone suggests that if educational stories (both creative nonfiction and fiction) are to reach their maturity, then some of them must be left at least momentarily unaccompanied by critique or theory. In the spaces created by holding back on theory lies the potential for what he calls *emancipatory educational story sharing*.

Here, the reader is coaxed into participating in the imaginative construction of literary reality through carefully positioned blanks in the writing, which are pauses that the active reader must fill with provisional meaning gathered from outside the text. Consequently, the blanks in the writing are used by what Barone (2000) calls the "artful writer-persuader," who understands the necessity of relinquishing control over the interpretations placed on a story and allowing readers to interpret the text from their own unique vantage points (see also chapter 5).

For Angrosino (1998), artful writer-persuaders, by letting go of the authoritative voice of omniscient science and moving toward fictional forms, are more able to create a world in which readers can interact with people and come to their own conclusions about what is going on: "The reader can do what the ethnographer does—immerse him- or herself in the particulars and try to figure out what it all means. And the reader might or might not come to the same conclusions as the ethnographer" (p. 41). Angrosino points out that in traditional scientific expository prose readers may have a sense that the author hasn't come up with a credible interpretation, but they don't have sufficient information to challenge the voice of authority, whereas "in the fictional genres, where life is acted out (shown, rather than told about), the reader is in a better position to draw his or her own conclusions" (p. 41).

Therefore, in this genre the relation between author, text, and reader is revised (see chapter 5). As Bochner notes, "the reader is repositioned away from being a passive 'receiver' of knowledge and elevated to the status of coparticipant in the creation of meaning" (1994, p. 31). In creating meaning from these different vantage points, readers can locate themselves in fictional tales and begin to perceive, experience, and understand what they have previously neglected. As Coles reminds us, "the beauty of a good story is its openness—the way you or I or anyone reading it can take it in, and use it for ourselves" (1989, p. 47).

Ethnographic fictions and creative fictions, by appealing to narrative ways of knowing and metaphorical truths, allow authors to represent absent, silenced, and marginalized others in ways that act as political points of resistance to their enforced absence, silence, and marginalization. This is so even when authors do not belong to the groups that they attempt to represent. Discussing whether it is presumptuous for a feminist writer to create characters quite different from herself, Livia comments on the case of Maureen Brady, a white lesbian writer who imagined the thoughts of the black characters in her 1982 novel, *Folly*:

> I asked a black Jewish friend what she thought of this issue. She agreed with Audre Lorde: the main problem facing black lesbians in literature (let alone black Jewish lesbians) is an almost complete absence. At least if a black lesbian were in there *somewhere*, my friend said, she could quarrel with the portrayal, start a conversation. . . . In whose interest is it that white women should feel black experience is so different from ours as to be unimaginable? Or that sighted women should believe the thoughts of blind women to be on such a different pattern that we will not venture to guess them? When gentile women read the work of Jewish women, it must be because we assume we will understand some, I venture to say even most, of what has been written. Probably we will not understand as much as other Jewish women, but enough to feel the book is worth our engagement.
>
> *Livia, 1996, p. 32*

Evoking the emotional texture of felt human experience; providing an audience with more immediate and authentic contact with others different to ourselves; reinforcing the shared subjectivity of experience; hearing the heartbeats of others; illuminating the quest for meaning; living outside of ourselves; perceiving, experiencing, and understanding what has previously been neglected; reorganizing experiences of the familiar; challenging habitual responses to the commonplace; offering new ways of perceiving and interpreting their significance; and "breathing together" would all seem worthy goals for researchers to aspire to in their writing. Ethnographic fiction and other kinds of story that condense, exemplify, and evoke a world are as useful for transmitting cultural understanding as any other researcher-produced device.

Good stories, according to Barone (2000), rattle commonplace assumptions; they surprise and disturb taken-for-granted beliefs to generate thought and discussion by suggesting rather than concluding. It would seem silly not to explore their potential in the future as a form of representation in sport and physical activity. Of course, this is not to imply that ethnographic fiction or creative fictions ought to substitute for other forms of ethnographic writing. Rather, as Angrosino argues, "it might profitably be considered a fully reputable menu item in the cafeteria of representations available to contemporary ethnographers, particularly those interested in exploring the ramifications of meaning in social behavior" (1998, p. 103). As one form among a variety of options, when chosen wisely with specific purposes in mind, ethnographic fictions and creative fictions have a great deal to offer.

Some Risks of Fictional Representations

There are some advantages to claiming that one's ethnographic writing is fiction. For example, L. Richardson (2000) notes that staging qualitative research as fiction frees the author from some constraints, protects the author from criminal or other charges, and may protect the identities of those studied. There are also disadvantages, however. For example, Banks and Banks (1998) believe that, despite a few exceptions, "fiction remains a no-no, a mode of expression—whether in film, writing, photography, or otherwise—that is simply off-limits in conventional academic discourse" (p. 17). Therefore, one obvious risk identified by Nilges (2001) in relation to what she calls postmodern textualization is "the possibility that research submitted for publication will be suspect. Most reviewers of professional journals in human movement and sport have been trained rhetorically to expect certain things from conventional texts not found in postmodern texts" (pp. 253-254). Producers of ethnographic fiction and creative fiction may find it more difficult to find outlets for their work in mainstream academic journals than those who produce the more traditional, and more accepted, realist tale.

In terms of literary skills, L. Richardson points out that "competing in the world of 'literary fiction' is very difficult. Few succeed" (2000, p. 933). To this, K. Frank adds, "It doesn't just happen, and it may be more difficult (emotionally or artistically) for the author than a traditional analysis" (2000, p. 485). Supporting this view, Denison and Rinehart (2000) emphasize just how tough it is to produce storied accounts that are able to contribute to our understanding of social life while also being artistically shaped and satisfying. For them, this "requires a high level of skill and dedication both to the craft of writing and the analytical skills of the scholar" (p. 3).

According to Sanders, although many genre-breaking stylistic experiments are noble and may result in strikingly unconventional materials when they are successful, researchers "frequently stumble and produce materials that read like high-school creative writing exercises or passages from mediocre cyberpunk novels" (1995, p. 95). Therefore, writing a good story, be it ethnographic fiction or creative fiction—just like producing a good ethnodrama, poetic representation, or realist tale—is no easy task, and the difficulties in terms of the literary skills called on should not be underestimated.

A further disadvantage is noted by L. Richardson when she states, "if one's desire is to effect social change through one's research, fiction is a rhetorically poor strategy. Policy makers prefer materials that claim to be not 'nonfiction' even, but 'true research'" (2000, p. 933). Indeed, tagging the word "fiction" to one's work creates the danger not only that policy makers will not take the work seriously but also that one's colleagues will have strong reservations about the research. If, as Banks and Banks (1998) suggest, fiction is a no-no in many disciplines, then, in a climate dominated by scientific and realist tales, some colleagues may suspect that those who produce creative fictions are not doing "proper" research at all but are frustrated novelists working in the wrong department! The concern is that the research is not founded, in a conventional sense, on data derived from direct involvement with actual people doing actual things together. For Sanders, the idea that research is not based on fieldwork is worrying:

> Fieldwork is an inconvenient, risky, dirty, and uncertain enterprise typically requiring considerable discipline and courage on the part of the investigator. What passes as post-modern ethnography often appears to be done by those who are shying away from the trials, tribulations, and pleasures of doing the "grunt work" of data collection. The result is a collection of seemingly rather lazy, couch potato, MTVish, ethnography-like works that are largely hermeneutic rather than rigorously empirical.
>
> *Sanders, 1995, p. 95*

In this regard, as part of his critique of postmodernism, Best suggests that it is "indoor work that requires no heavy lifting" (1995, p. 127). By this he means that little sociological work is required in the form of fieldwork. Thus, he sees recent calls for textual experimentation as calling for a move "out of the streets, into the armchairs" (p. 128).

Needless to say, those who produce creative fictions can expect a more hostile response from their traditionally inclined colleagues than from those who produce ethnographic fictions. At least, so the argument goes, the latter have actually done some fieldwork and collected some data. Even this, however, does not guarantee a trouble-free ride. The demeaning of one's work can have a powerful impact on one's identity and feelings of self-worth as well as having a direct influence on career prospects and promotion opportunities. For L. Richardson (1993, 1996a, 1997), any text that violates conventions by stepping outside the normative constraints of social science writing is vulnerable to dismissal and to trivialization as commonplace. This point is echoed by Pelias (1999), who argues that writers of poetic essays (that is, ethnographic fiction and creative fictions) risk the charge of irrelevance because they seldom specify how their contributions add to the ongoing knowledge in the field.

> It is common in the traditional essay for a writer to first identify what has been done on a given topic and then to articulate how his/ her essay will explore new terrain. Such markings are typically not done in the poetic essay, since to do so would be an acceptance of the positivist presupposition that knowledge is progressive, always moving toward the goal of obtaining the complete truth. This assumption is one that writers of the poetic essay would reject.
>
> *Pelias, 1999, p. xiv*

Pelias (1999) also points out that writers of poetic essays risk further accusations of irrelevance, and readers may have difficulty placing the work in a scholarly context, because "traditional procedures, such as reviewing the literature, citing sources, and building bibliographies, may be left behind" (p. xiv). Accordingly, Van Maanen expresses his concern that writers of literary tales seem content to allow their accounts to stand alone with little or no mention of previous work in the same area or a similar one: "The representations are typically cast as discoveries and come forth in something of a scholarly vacuum" (1988, pp. 134-135). To this, he adds that these tales "can be fluff—merely zippy prose on inconsequential topics. They are liable then to the charge of 'scoop ethnography'" (p. 135).

This charge is particularly relevant given ethnographic fiction's need to maintain the reader's interest. At times, the need for a good story can override the need for a good analysis. In this regard, Van Maanen suggests that literary tales might be so tied to the representational techniques of realist fiction that they actually distort the very reality they seek to capture: "The need for action, drama, high-jinx, colorful characters, and purple prose may drive out the calmer, more subtle and sublime features of the studied scene" (1988, p. 135). Of course, he adds, "Such flaws are not inherently genre specific" (p. 136).

Sandelowski (1994) also expresses the view that celebrating the art in qualitative research does not give us the license to fail to know about or acknowledge the work of other people in our own and other disciplines when presenting our work, or to fail to compare our findings with relevant ones presented by other researchers, no matter what their research orientation might be.

Qualitative research is not a license to reinvent the wheel in the service of some misguided notion of bracketing one's assumptions or maintaining a pristine and pure theoretical stance. Such putative innocence of other people's work makes the researcher guilty of breaking the cardinal rules of scholarship; Know thy field and give credit where credit is due.

Sandelowski, 1994, p. 59

Finally, there remain tensions and concerns in the qualitative research community even among those who support the use of alternative forms of representation; these concerns revolve around the fact/fiction dilemma and the issue of *making it up*. Here, as Rinehart (1998a) acknowledges, "From the point of view of the academic ethnography writers, fictional ethnography contains falsehoods and is often more flair than substance. Conversely, from the point of view of fictional ethnographers, some of the questions academic ethnographers ask have become mere listings" (pp. 204-205).

Accordingly, Van Maanen (1988) expresses concerns regarding what he calls literary tales. These include the concern that there is often either too much or too little of the author in the tale. There is no way of knowing if the author really got it "right." There is too much unlicensed interpretive work in the text, so that the author's credibility is open to question. Plus, there is always the charge that *"The bastards are making it up!"* On this issue, Sandelowski comments, "Qualitative researchers are not free to make wild forays into fancy; they make, but cannot fake" (1994, p. 58). This stance is supported by several of those who strongly advocate innovative writing practices. For example, Bochner and Ellis emphasize the need to appreciate the differences between *making* something and *making it up*.

The idea of blurring genres of inquiry may help obscure the boundaries between science and literature, but it doesn't obliterate the responsibility to try to be faithful to our experiences in the field. . . . that's the danger of going too far with the notion of ethnographic fiction. We ought to treat our ethnographies as partial, situated, and selective productions, but this should not be seen as a license to exclude details that don't fit with the story we want to tell. I want to retain a distinction between saying our work is selective, partial, and contestable, and saying that the impossibility of telling the whole truth means you can lie. If we unquestionably accept the depiction of the ethnographer as a trickster, a sophist, and a politician, we aren't far from seeing her as a liar.

Bochner & Ellis, 1996b, pp. 21-22

L. Richardson (1997) makes a similar observation. Having acknowledged that all writing—scientific and literary—depends on literary devices not only for adornment but also for conveying content, she warns against the argument that all ethnographic writing is "fiction," because no "facts" ever exist in and of themselves but rather exist only as interpreted facts. For Richardson, if ethnography claims to be *only* "fiction," then it loses any claims it might have for groundedness and policy implications. She suggests the direction for ethnography "is not to deny its social scientific grounding but to take this opportunity to explore its grounds for authority, partial and limited as they may be" (p. 108).

That ethnographers utilize fictional techniques to represent their findings in the form of creative nonfiction seems less worrisome to many than the notion of researchers using their imaginations to invent a story. Being there, as opposed to not being there, remains a core concern. In combination, the concerns expressed in this chapter may, at times, outweigh the undoubted benefits of using this genre for certain purposes in certain contexts. Therefore, even though well-crafted ethnographic and creative fictions certainly have much to offer, choosing these genres takes authors into highly contested terrain. They need to be prepared.

Reflections

The emergence of fiction in the domain of sport and physical activity can be located as an attempt by some scholars to respond constructively to the recent challenges posed by postmodernist and poststructuralist perspectives on truth, neutrality, objectivity, and language (see Nilges, 2001; Denison & Markula, 2002; Sparkes, 1991, 1995, 1998c, 2001). Of course, this is not the only response, or the "correct" one, but it is a worthy response.

As part of this response, in keeping with other disciplines in the social sciences, there is a move from *description* to *communication* (Bochner & Waugh, 1995). On this issue, Bochner and Ellis (1996a) speak of an interactive ethnography that privileges the way in which researchers are part of the world they investigate and the ways in which they make it and change it. This breaks away from the epistemology of depiction that privileges modes for inscribing a preexisting and stable world:

For writers whose work departs from canonical forms of narrating ethnography, there is a desire to be more author centered and, at the same time, more engaging to readers. Forms and modes of writing become part and parcel of ethnographic "method." The goal is not only to know but also to feel ethnographic "truth" and thus to become more fully immersed—morally, aesthetically, emotionally, and intellectually. We think that the stories ethnographers will tell as they embrace these goals and challenges will be the kind that activate subjectivity and compel emotional response. These stories will long to be used as well as analyzed, to be revised and retold rather than settled and theorized, and to promise the companionship of concrete, intimate detail as a substitute for the loneliness of abstracted facts.

Bochner & Ellis, 1996a, pp. 4-5

Reflecting on one of his responses as an ethnographer to the challenges of postmodernism, Gottschalk (1998) also notes a shift from authoritative "describing" to a more self-reflexive "inscribing" and a shift from the pretentious "representing" to a more modest "evoking." Rather than attempting to convince readers of the truth of their accounts by appealing to traditional and increasingly challenged authorities and criteria, postmodern ethnographers seek instead to promote an understanding through "recognition, identification, personal experience, emotion, insight, and communicative formats which engage the reader on planes other than the rational one alone. They seek to evoke the postmodern culture, moment, and consciousness rather than to describe it" (pp. 213-214).

For many scholars, as I have tried to illustrate in this chapter, the move from description to communication is provocative, risky, unsettling, and threatening. This move raises issues and questions for debate. Indeed, we might begin by reflecting on why it is that the move from description to communication is so threatening to some. Why is it that some researchers in the field of sport and physical activity feel so threatened by fictional representations? Why is it that fictional representations seem to pose such a danger and enrage so many? Why is it that many proceed with undue haste to define fictional representations, along with other experimental genres (e.g., autoethnography, ethnodrama, and poetic forms), as "not research" or "not sociology or psychology"?

Of course, this is not to suggest that producers of ethnographic fiction, creative fiction, or any other experimental genre should not also be asked some difficult questions. Given that no textual staging is innocent, their work, like any other form of scholarship, should not be accepted uncritically. For example, drawing on Kress (1998) and Rath (2001), we might ask, While qualitative research into sport and physical activity can, and does, benefit from the use of fictional techniques, does the crossing of the boundary (however artificial and contrived) from ethnography to fiction without bringing any sociological/psychological baggage mean we lose (whatever losing means) too much in the process? If sociology or psychology is what sociologists or psychologists *do*, does this dismiss all boundaries and abandon all definitions? What problems arise when boundaries are disregarded?

Furthermore, is it possible that boundaries are useful precisely because they make play possible? Can there be something called "interdisciplinary" without the existence of disciplines? Can there be creativity without constraints? If a piece of fiction can be good qualitative research, does it matter if it is *good* fiction? And, if a piece of fiction does what, say, sport sociology or sport psychology does, why do we need these disciplines? How can fiction and social research comfortably cohabit? What new forms of social research may be generated through an intimate relationship with fiction? How might researchers have to change their relationship to their work, in terms of how they know and tell about the sociological or psychological, in order to conjure up a different kind of social science? Of course, we might also ask, When is a piece of fiction or a lyric poem *not* ethnographic or sociological or psychological?

Addressing such questions provides an opportunity for dialogue and debate both within disciplines and across disciplinary boundaries. For example, if, for the sake of argument, we momentarily accept that a piece of fiction can be good qualitative research in sport and physical activity, then I think that it should be "good" rather than "bad" fiction. But this raises the question as to how might we judge a piece of writing as good fiction. More traditional criteria—like validity, reliability, generalizability, and other standards associated with correspondence notions of truth—do not seem appropriate for judging ethnographic fiction, creative fiction, and other experimental genres. So, just what criteria might we call on to pass judgment? The whole of chapter 9 is devoted to exploring this crucial question.

In the meantime, those who have taken the risk and produced well-crafted ethnographic fictions and creative fictions as a form of analysis that evokes the emotions—enabling a wide range of audiences to viscerally inhabit and understand different worlds in ways that convey complexity and ambiguity without producing closure—have provided a valuable intellectual stimulus to qualitative researchers in sport and physical activity. I, for one, am grateful to these scholars for the courage they have displayed in taking the risk.

Chapter 9

Different Tales and Judgment Calls

There are now many ways that qualitative researchers in sport and physical activity can represent their findings, ranging from the more traditional realist tale to ethnographic fictions and creative fictions. As Denzin argues, there is more than one way to do representation: "There are several styles of qualitative writing, several different ways of describing, inscribing and interpreting reality" (1994, p. 506). Indeed, the recent and daring explorations in new modes of research remind Eisner of the Cole Porter lyric: "In olden days a hint of stocking was looked on as something shocking, now heaven knows, anything goes. Good authors too who once knew better words now only use four-letter words—writing prose. Anything goes" (2001, p. 139).

Of course, as the previous chapters in this volume make clear, simply writing in a different genre does not necessarily ensure a better product. There can be good and bad autoethnographies just as there can be good and bad realist tales. That is, "not anything goes." But just what do we mean by good and bad in relation to different

tales? How are we to judge new ways of writing? What criteria should we use? As DeVault's comment regarding autoethnography suggests, judging different tales takes many scholars into new territory:

> Sociologists are unaccustomed to evaluating personal writing; the standards for critique and discussion seem "slippery" to many in comparison with more familiar criteria associated with the scientific research report. . . . As personal writing becomes more common among social scientists, researchers will need to develop new avenues of criticism and praise for such work. . . . Presumably, a "good" story in some contexts, for some purposes, may not be so good for others. Such criteria for evaluating personal writing as sociology have barely begun to develop.
>
> *DeVault, 1997, p. 24*

Questions of judgment regarding new writing practices in qualitative research have been raised by a number of leading scholars and lie at the heart of the legitimation crisis described by Denzin & Lincoln (1994, 2000; see also chapter 1 of this book). For example, Denzin asks, "Is any representation of an experience as good as any other?" (1997, p. 5). Talking of the new frontiers in qualitative research methodology, Eisner (1997) asks, "How should work in the new frontier be assessed? What criteria seem appropriate?" (p. 268). How can we separate the wheat from the chaff? Lieblich, Tuval-Mashiach, and Zilber ask, "What then can be offered as criteria for the quality of narrative research? How should we distinguish a good study from a bad one?" (1998, p. 171). Josselson asks, "What is a good story? Is just a good story enough? What must be added to story to make it scholarship?" (1993, p. xi). Similarly, L. Richardson and Lockridge ask, "if one chooses to use fictional techniques in the writing of ethnography, how are those ethnographies evaluated? How might one judge them? What might be the criteria?" (1998, p. 328).

Judgment Calls: A Personal Example

According to Anderson, journal editors and reviewers for qualitative publications now see an assortment of texts based on a variety of epistemological and methodological assumptions, "and they face the difficult task of evaluating the manuscripts according to sometimes incommensurable criteria" (1999, p. 453). Here, my own experiences might shed some light on the outcomes of this difficult task. As elsewhere (Sparkes, 2000), I have focused specifically on the criteria used by six reviewers of a leading journal to judge my own autoethnography, "The Fatal Flaw" (see Sparkes, 1996).

These ranged from very positive to very negative comments. For example, one reviewer commented,

> The first part of this paper is breath-taking. Sparkes situates us in poststructuralism—brings us into our own body understanding of how it is to live in a poststructuralist world. Then, in an incredibly compelling story/complex narrative construction of the loss of the body/self, he accomplishes what he wants to happen: we read ourselves (I read myself) as a body; frail, too.
>
> *Sparkes, 2000, p. 25*

In contrast, another reviewer stated the following:

> Let me preface my comments by saying that in general I'm not positively predisposed to studies based on a narrative of one person. . . . It is very hard to make good sociology—which is what I think this paper wants to be—from a single case study, especially if it is one's own. We all have stories. Indeed, we all have lived lives, and I'm not sure we're doing scholarship, and sociology, a favor by "sociologizing" them.
>
> *Sparkes, 2000, p. 27*

These two reviewers are passing judgment on my work, but they seem to be using different criteria to do so. The negative comments by the latter are not uncommon, and they represent the majority view toward the telling of different tales in qualitative inquiry. This view is the voice of traditional science that is committed to "rationality," "objectivity," and a range of dualisms that include subject/other. It holds to certain epistemological and ontological assumptions and calls on specific foundational criteria to pass judgments on research. Numerous critiques are now available of this position and the dangers of imposing inappropriate judgment criteria on different forms of inquiry, so I will not rehearse them here (see Guba, 1990; J. Smith, 1989, 1993). Needless to say, when standard, traditional criteria of what makes a good "scientific" or "realist" telling are applied, then confessional tales, autoethnographies, poetic representations, ethnodramas, and ethnographic fictions will always disappoint.

This point is emphasized by Garratt and Hodkinson (1998), who focus on the published version of "The Fatal Flaw." Given the unusual nature of this article, they ask, how might it be possible to begin making judgments about its quality and worth? Garratt and Hodkinson then illustrate the dangers of applying *inappropriate* foundational criteria, such as plausibility and credibility, as advocated by the neorealist Hammersley (1992). With regard to plausibility, this approach would entail asking ourselves whether the claims made in the research seem plausible given our existing knowledge. As Garratt and Hodkinson point out, however:

> How could we begin to make judgments about an autobiographical narrative of the self, immanently characterized in terms of its subjectivity, uniqueness, fragmentation, and novelty of expression, on the basis of either the empirical claims it makes or in terms of the match with existing research. . . . In this sense, it may be neither possible nor desirable to judge the quality of this piece on the basis of its accuracy relative to research that already exists within the field. Rather, it might be more helpful, in this particular case, to ask ourselves different questions, such as: Does this account work for us? Do we find it to be believable and evocative on the basis of our own experiences?
>
> *Garratt & Hodkinson, 1998, pp. 252-253*

Given that "The Fatal Flaw" is *bound to fail* the test of plausibility in Hammersley's (1992) terms, the next step is to assess its credibility. Again, Garratt and Hodkinson (1998) point out the problems with such a move. As they comment, any judgment about the credibility of a claim necessarily involves a judgment about its accuracy, which entails a closer examination of the evidence collected and a judgment about whether the research was conducted using the relevant methodology. Garratt and Hodkinson note that such procedures appear incongruous in understanding the research process associated with the production of "The Fatal Flaw," where I actually state that accuracy is not the issue, since autoethnography and narratives of the self seek to meet literary criteria of coherence, verisimilitude, and interest.

> In which case any judgment about the value of research would rest primarily on feelings of trust and the experience of the reader in participating with the text. To attempt to judge this work against Hammersley's (1992, 1995) criteria of plausibility and credibility will inevitably result in the work being described as seriously flawed or classed as *not research*.
>
> *Garratt & Hodkinson, 1998, p. 526*

If alternative forms of qualitative inquiry and new writing practices are judged using inappropriate criteria, there is the danger that they will be dismissed as not being proper research and, therefore, not worthy of attention. For example, Schwalbe argues that whatever the value of poetry as literary therapy, "it is not an adequate means to meet our aims and responsibilities as sociological analysts" (1995, p. 406). He adds, "perhaps the turn to poetry and other forms of experimental writing is mostly a way to keep ourselves awake and amused before retirement" (p. 411). As Denzin points out, however, "Predictably, and regrettably, his [Schwalbe's] criticisms fail to engage the new writing on its own terms. Consequently, the grounds for a fruitful conversation are not created" (1996, p. 526; see also DeVault, 1996; L. Richardson, 1996b; St. Pierre, 1996). Indeed, sometimes when the new is derided and treated with open hostility, there is no room for conversation at all.

Experiencing Transgression

As a pioneer of experimental forms of representation, and as one who tries to write sociology that moves people emotionally and intellectually, L. Richardson (1993, 1996a, 1997) provides some revealing insights into how it *feels* to transgress boundaries and have your work judged by inappropriate criteria. In "Educational Birds" (1996a), she uses a dramatic representation to focus on the experiences of an academic who moves into contested terrain by daring to write differently. Following are several scenes from this drama.

Scene 1

It is a chilly September afternoon in a sociology department chair's office. The walls are catacomb yellow; there are no mementos, pictures, or plants in the room. Seated at one end of a large conference table are two women: the department chair, with her back to the windows, and Full Professor Z looking out to the silent gray day.

Chair: I've been reading your work, because of the salary reviews—

Professor Z: —

Chair: —You write very well,

Professor Z: —

Chair: But is it sociology?

Professor Z: —

(p. 6)

Scene 3

It is an overcast November noon at the faculty club. Pictures of the deceased faculty, men in drab suits, line the room; wrought-iron bars secure the windows. Professor Z and Assistant Professor Q, whose five-authored paper "Longitudinal Effects of East to Midwest Migration on Employment Outcomes: A Log-Linear Analysis" has made her a member of the salary committee, are eating the "crimson and steel" lunch specials.

Assistant Professor Q: Everyone says, "You write very well."

Professor Z: Is that a compliment?

Assistant Professor Q: But is it sociology?

(p. 7)

Scene 6

A chilly March evening in a conference room in a convention hotel in Metro-City. The room has no windows, one rear door, and an erratic heating system. No one has been outside all day. Professor Z is at the podium.

Professor Z: Why prose? Prose, I submit, is not the only way to represent sociological understanding. Another possible way is through poetic representation. Poetry touches us where we live, in our bodies, and invites us to experience reflexivity and the transformational process of self-creation.

For this paper, I transform an in-depth interview into a narrative poem.

Using only the interviewee's words, diction, rhythms, and other poetic speech components, I try to meet both scientific and poetic criteria. *(Professor Z reads poem and discusses Derrida, Denzin, and data, poetically represented, which is followed by a discussion period.)*

Conferee 1: Where is the f—king validity?

Conferee 2: What about reliability?

Conferee 3: Truth? Where's truth?

Conferee 4: And reality?

Conferee 5: Have you lost your f—king mind!!!

Professor Z: *(Takes field notes)*

(p. 8)

In a later scene, Professor Z attends a conference where a flock of sociologists proceed to define their own particular brand of research as "sociology" but feel it is inappropriate to pass poems off as "sociology."

In another article, L. Richardson (1993) recounts in dramatic form her own experiences of transforming an open-ended life history interview into a five-page poem and then presenting this poem at a major conference. Once again, questions of validity, reliability, truth, and the status of the interview "data" are raised by the audience of sociologists. In her interpretation of the reactions to her poem, Richardson notes that turning interview data and writing a life as a poem display how sociological authority is constructed and thereby problematize reliability, validity, and truth. Furthermore, she argues, "Poetics strips these methodological bogeymen of their power to control and constrain. A poem as 'findings' resituates ideas of validity and reliability from 'knowing' to 'telling.' Everybody's writing is suspect—not just those who write poems" (p. 706). Elsewhere, Richardson (1997) acknowledges that her poetic representation, as a text that violates conventions and destabilizes traditional forms of judgment, was defined as an act of heresy in sociological terms by her audience at the conference because it violated their sense of safety and security. Therefore, presenting the unexpected can be problematic.

Perhaps, as L. Richardson (1996b) suggests, the reactions she encountered at the conference and from her colleagues occurred because her novel approach to presenting data and her attempts to write differently in general breach sociology's representational rules. This breaching challenges the taken-for-granted codes through which members have been socialized into role taking and "consciousness of kind," and through which academic communities have exerted social control over their members. As a direct result, members often respond to breaches through behaviors that indicate anger or dismissiveness. In "Educational Birds," Professor Z ruminates on these issues and offers the following opinion:

> I know why these people feel so threatened. They fear that if any rule is violated, all rules might be violated. They fear lack of control not only in their professional but in their personal worlds. The subtext of the question "But is it sociology?" is their silenced fear: "If poetry is sociology and I can't do it, what happens to *my* identity, *my* prestige, *my* status—*my place in the pecking order—ME . . . Me, me. . . .*
>
> L. Richardson, 1996a, p. 13

The views expressed by Professor Z may seem harsh and rather unkind to many scholars. As I have argued elsewhere (Sparkes, 1991, 1995), however, power, politics, and regimes of truth *do* operate in academic communities. These serve the interests of specific groups by legitimating some forms of inquiry and representation at the expense of others, by including and excluding, and by foregrounding certain voices and silencing others. The effects of this power to define are very real in terms of publication outlets, employment, grants, promotion, and prestige.

Young scholars learn early in their careers to understand what is expected of them. As Nilges (2001) reflects, in thinking about the publication of her first qualitative research study, "I knew I had made many decisions relating to textual forms . . . with the hope that it would not only present a convincing argument for rethinking gender equity in physical education but that it would be viewed *as creditable by the research community*" (p. 231; my italics). After reviewing the contents and style for papers produced in two leading journals over a period of several years, Nilges came to understand that qualitative research was written "within very narrowly prescribed boundaries, even though the theoretical and analytical approaches motivating such work differed greatly" (p. 232). She goes on to state the following:

> I knew that if my paper was to have a chance of being published, there was an "acceptable" style of qualitative writing that I needed to follow. . . . This style of writing was not foreign to me. During graduate school, my professors stressed the importance of conventional textualization where credibility is established through the use of neutral language, a detached (omniscient) author, neatly packaged themes, and a clear separation of methods, results and recommendations. An "acceptable" thematic piece was eventually written and published.
>
> *Nilges, 2001, p. 232*

Nilges (2001) recognizes the conservative climate in which she operates. So do many others. This is a climate in which anything "new" is viewed with suspicion, if not hostility—a climate in which moves are made by those in power to keep the academic game the same so that the status quo remains undisturbed. There is a tendency to construct and preserve tight symbolic membranes between different approaches in any given discipline, and to constrain the complexities of social inquiry in the boundaries of narrowly sectarian theory or entrenched positions. These attitudes and tactics might benefit certain individuals and groups by encouraging compliance to accepted ways of doing things. It is questionable, however, whether they benefit the development of the discipline as a whole by encouraging what I have called elsewhere (Sparkes, 1991) a polyvocal research community that is spoken, written, and represented from many sites.

Dealing With Difference

Any kind of research can be dismissed, trashed, and trivialized if inappropriate criteria are imposed on it. Being aware of this situation goes some way to resolving the dilemmas highlighted by Foley, who, in relation to his own attempt to write a critical sports narrative, asks:

> How is it possible to write truly accessible, popular representations of sports that are also reflexive and thus fulfill the criteria of good postpositivist critical interpretation? . . . One is left feeling that there is no way to serve two masters, the people and the professorate. . . . I am still trying to find a middle road between extensive poetic experimentation advocated by some postmodern ethnographer's . . . and the new, more accessible quasi-literary version of the old scientific realist tale. . . . Is it possible to write a thoroughly popular, dramatic, accessible narrative that social scientists will still consider scientific?
>
> *Foley, 1992, pp. 44-45*

The simple answer is a resounding "No!" As J. Smith (1993) emphasizes, since different epistemological and ontological assumptions inform qualitative and postpositivistic inquiry, it makes little sense to impose the criteria used to pass judgment on one on the other. Likewise, Plummer (2001) argues that conventional modes of evaluation may not be suitable for judging life history research. For him, "different goals and different kinds of data require different modes of evaluation" (p. 153).

Indeed, as I have argued elsewhere (Sparkes, 1992, 1995, 1998c, 2001), attempts to impose inappropriate criteria on work that is different from one's own is at best misguided and, at worst, arrogant and nonsensical, a form of intellectual imperialism that builds in failure from the start so that the legitimacy of other research forms is systematically denied. As a consequence, the research community is left in a "no win" situation in which researchers offer blind allegiance to their own particular paradigmatic positions and refuse to acknowledge the contribution that other ways of knowing can make to our understanding. In view of this, I have suggested, the differences between alternative forms of inquiry, in terms of their process and products, need to be acknowledged so that each can be judged using criteria that are consistent with their own internal meaning structures and purposes. This view is supported by Markula, Grant, and Denison, who, in a review of research into aging and physical activity, state the following:

> Because we no longer have a unified research philosophy, neither can we demand a unified criterion that validates all research. Instead

of such a criterion, we need to better understand the premises of each research tradition to ensure that their results contribute to our knowledge. . . . It is inappropriate to judge a poststructuralist deconstruction of the word *aging* based on positivist criteria or, vice versa, to demand that a positivistic study include a rich description of an individual's exercise experience. Instead we need to judge the meaning of each research project against its philosophical premise. . . . The acknowledgement of multiple research traditions does not translate into an acceptance of poor-quality research but instead requires us to contextualize each study carefully within its paradigm.

Markula et al., 2001, pp. 261-262

Of course, acknowledging difference in the work of others and judging their work fairly are not easy matters. As Denison and Rinehart point out,

To us the shift into personal experience narratives, storied accounts, or ethnographic fiction requires a highly complex and complicated conceptual shift in the way one approaches subjects and topics. It requires a shift in sensibilities—perhaps even in world view—that we believe should affect every decision the researcher makes about the study all the way from the germ of an idea to the field to the text.

Denison & Rinehart, 2000, p. 3

Furthermore, the realization of difference creates an ethical imperative in which the task, according to Bernstein (1991), is to listen carefully and attempt to grasp what is being expressed and said in alien traditions. For him, this needs to be done in ways that resist the dual temptation of either facilely assimilating what others are saying into our own categories and language without doing justice to what is genuinely different or else simply dismissing what the other is saying as incoherent nonsense. According to him, such responsibility "should not be confused with an indifferent superficial tolerance where no effort is made to understand and engage with the incommensurable otherness of the 'Other'" (p. 66).

In this regard, Sandelowski reminds us, "like any other serious project, a qualitative piece indeed may be judged as good or bad in terms of certain cannons of inquiry and aesthetics in the same way that a novel, drama, or conventional scientific product may be judged" (1994, p. 60). For her, the key issue is the need to educate ourselves (as practitioners, critics, and consumers of research) to recognize the difference and judge the genres accordingly using appropriate criteria. Eisner makes a similar point when he argues that "if *how* things are said is relevant for understanding the meanings of a message, one needs to be able to 'read' these meanings and understand the forms that convey them" (1991, p. 51). In his terms, we need to become *connoisseurs* of different forms of qualitative inquiry and the associated new writing practices so that we appreciate the contribution they make to our understanding.

Validities, Truths, and Different Ways of Knowing

The need to use appropriate but different criteria to judge various forms of representation is evident when we consider issues of truth, or validity. This is particularly so given that different tales construct various kinds of truth in different ways. For example, Banks and Banks suggest, "Pablo Picasso said that we all know art is not truth. Art is a lie that makes us realize truth. The creator of fiction must know how to convince others of the truthfulness of his or her lies" (1998, p. 17). They go on to argue that "The standards of adequacy for academic productions must be revised once fiction is admitted to the dance" (p. 18). Likewise, in talking about his own poetic essays, Pelias (1999) notes, "They are fictions that never assert the truth but strive to be truthful. . . . They are imaginative constructions whose truth lies not in their facticity but in their evocative potentiality" (p. xiv).

When it comes to telling alternative tales, orthodox "scientific" views of validity (and reliability and generalizability), based on positivistic epistemological assumptions that adhere to correspondence notions of truth, make little sense. For example, R. Atkinson points out that "reliability and validity are not necessarily the appropriate evaluative standards for a life history interview" (1998, p. 59). Likewise, Riessman notes that the prevailing procedures for establishing validity "rely on realist assumptions and consequently are largely irrelevant to narrative studies. A personal narrative is not meant to be read as an exact record of what happened nor is it a mirror of the world 'out there'" (1993, p. 64). For her, validation in narrative studies cannot be reduced to a set of standardized technical procedures, so that "traditional notions of reliability simply do not apply . . . and validity must be radically reconceptualized" (p. 65).

One move, within what I have defined as a *diversification perspective* (Sparkes, 1998c), is to radically reconceptualize or reframe the notion of validity to judge different forms of inquiry. For example, for those qualitative researchers whose work is informed by a critical agenda that is openly ideological, disclaims any notion of value neutrality, and aspires to emancipate those involved in the research process by empowering them to take control of their own lives and challenge the status quo, the issue of validity takes on various meanings.

Writing from a feminist perspective, Lather (1986) speaks of *catalytic validity.* This term refers to the degree to which the research process energizes participants and alters their consciousness so that they know reality and can better transform it. A similar point is made by Krane (1994) in her review of a feminist perspective on contemporary sport psychology research, when she notes how this kind of research can be enlightening and empowering for women surrounded by male interpretations of sport and exercise. For her, one intent of feminist theory and research is to provide solutions that eventually lead to social change. Here, feminist perspectives are taken to reflect a commitment to social change and increased justice; these perspectives are to be contrasted with traditional positivistic science, which usually employs a structuralist-functionalist perspective that seeks to maintain the existing social order.

Similar issues are raised by McTaggart (1997) in relation to participatory action research. He suggests that validity, in the context of this form of inquiry, needs to be reconceptualized in terms of the efficacy of the research in changing relevant social practices: "Our thinking about validity must engage more than mere knowledge claims. . . . [It must be] comprehensive enough to reflect [what] the social action researchers . . . are committed to, by the trenchant and withering critiques of other more inert and detached forms of social inquiry" (pp. 17-18).

Validity in critical and participatory action research, therefore, might involve some evaluation of how effective the research process and the research product (in the form of text, be it poetic, autoethnographic, performance, or fictional) have been in actually empowering the participants and enabling them to create change. That is, has the writing made a difference? Has it moved people to action?

Another move toward reconceptualizing validity is made by Lather (1993) as part of her own fertile obsession with the topic of validity after poststructuralism. For Lather, validity is an incitement to discourse. Her goal is to reinscribe validity so as to loosen the master code of positivism and postpositivism that she feels continues to shape contemporary research.

> Rather than jettisoning "validity" as the term of choice, I retain the term in order to both circulate and break with the signs that code it. What I mean by the term, then, is all of the baggage that it carries plus, in a doubled-movement, what it means to rupture validity as a regime of truth, to displace its historical inscription toward "doing the police in different voices."
>
> *Lather, 1993, p. 674*

In terms of redefining judgment criteria, Lather (1993, 1995) raises a host of concerns that decenters validity as being about epistemological guarantees and reframes it as multiple, partial, and endlessly deferred.

Accordingly, her notion of a *validity of transgression* runs counter to the standard validity of correspondence. In developing her ideas on transgressive validity, Lather discusses the strands of, and provides checklists for, *ironic* validity, *paralogical* validity, *rhizomatic* validity, and *voluptuous* validity. While elements of these strands are to be found in the alternative tales I have discussed in the previous chapters, the notion of voluptuous validity is particularly appealing. With regard to this form of validity, Lather notes the following features:

- goes too far toward disruptive excess, leaky, runaway, risky practice.
- embodies a situated, partial, positioned, explicit tentativeness.
- constructs authority via practices of engagement and self-reflexivity.
- creates a questioning text that is bounded and unbounded, closed and opened.
- brings ethics and epistemology together.

Lather, 1993, p. 686

The poetic representations of L. Richardson (1992a) are cited by Lather (1993) as an example of the kind of work that displays voluptuous validity and goes "too far" with the politics of uncertainty by blurring the lines between the genres of poetry and social science. This criterion, along with the other forms of validity outlined by Lather, might also be useful for judging some of the other tales discussed in previous chapters—for example, autoethnographies, ethnodrama, and fictional representations.

L. Richardson (2000) also offers a transgressive take on validity by reconceptualizing it in terms of the *crystalline*. That is, validity, in a metaphoric sense, has the properties of a crystal rather than the fixed points assumed in methods triangulation.

I propose that the central imaginary for "validity" for postmodern texts is not the triangle—a rigid, fixed, two-dimensional object. Rather, the central imaginary is the crystal which combines symmetry and substance with an infinite variety of shapes, substances, transmutations, multidimensionalities, and angles of approach. Crystals grow, change, alter, but are not amorphous. Crystals are prisms that reflect externalities *and* refract within themselves, creating different colors, patterns, and arrays, casting off in different directions. What we see depends upon our angle of repose. Not triangulation, crystallization.

Richardson, 2000, p. 934

The notion of crystallization suggests that we might draw on multiple ways of judging qualitative forms of inquiry and that these might vary over time, depending on the angle of repose, for the same piece of research.

From Validity to Verisimilitude

As part of what I have called a *letting go* perspective (Sparkes, 1998c), some researchers have chosen to abandon the notion of validity completely. For example, in discussing his own impressionist tale—which incorporates a narrative of self to provide dramatic and destabilizing insights into personal, legal, and medical (psychiatric) "truths" as they impact his life—Wolcott talks of the *absurdity of validity*.

> What I seek is something else, a quality that points more to identifying critical elements and wringing plausible interpretations from them, something one can pursue without becoming obsessed with finding the right or ultimate answer, the correct version, the Truth. . . . And I do not accept validity as a valid criterion for guiding or judging my work. I think we have labored far too long under the burden of this concept (are there others as well?) that might have been better left where it began, a not-quite-so-singular-or-precise criterion as I once believed it to be for matters related essentially to tests and measurement. I suggest we look elsewhere in our continuing search for and dialogue about criteria appropriate to qualitative researchers' approaches and purposes.
>
> *Wolcott, 1994, pp. 366-369*

A useful starting point for "looking elsewhere" might be the world of literature and literary criticism. This is particularly so if fictional constructions of experience are to become a legitimate research form. After all, standard tests like reliability, validity, and replicability are simply not appropriate for judging fiction. Such works are better judged by aesthetic standards, by their emotive force, by their capacity to engage readers emotionally, by their verisimilitude, and by their authenticity or integrity. In support of this view, Eisner argues that, with regard to nonpropositional language, the *rightness* of fiction is not the *truth* of physics:

> To say a work of fiction is "right" or "true to life" is to say that it captures and illuminates some aspect of reality. This reality is a function of the writer's ability to use a form we call literature to disclose something about the world that he or she has experienced and that we find believable. Scientific truth tests are as relevant to testing fictional truth as knowledge of chemistry is relevant to making soufflés.

To dismiss the ways in which literature or poetry inform because they cannot be scientifically tested is to make a category mistake. We can live with many versions of rightness, truth being one.

Eisner, 1991, p. 50

Not surprisingly, Rinehart (1998a) argues that for the human disciplines it might be better to borrow criteria from the arts and from aesthetics rather than from pure science. For him, we need to "establish multiple bases of evaluation, all equitably privileged. A broad model of evaluation is one based on pragmatism: Is the work effective?" (p. 124).

These suggestions become even more compelling when we consider the views of Bruner (1986) about different modes of thought. He proposes that there are two modes of cognitive functioning, or two modes of thought: The *paradigmatic*, or *logico-scientific*, and the *narrative* modes, each of which provides distinctive ways of ordering experience and constructing reality. The former attempts to fulfill the ideal of a formal, mathematical system of description and explanation and deals in general causes and their establishment; it makes use of procedures to assure verifiable reference and to test for empirical truth. In contrast, Bruner describes the *narrative* mode of knowing as follows:

> The imaginative application of the narrative mode of knowing leads instead to good stories, gripping drama, believable (though not necessarily "true") historical accounts. It deals in human or human-like intention and action and the vicissitudes and consequences that mark their course. It strives to put its timeless miracles into the particulars of experience, and to locate the experience in time and place.

Bruner, 1986, p. 13

Although these two modes can be *complementary*, they are *irreducible* to one another. Indeed, Bruner (1986) warns that to reduce one mode to the other, or to ignore one at the expense of the other, will inevitably fail to capture the rich diversity of thought contained in each. Quite simply, each is a *different* way of knowing, with its own operating principles and its own criteria of well-formedness. As a consequence, the two modes differ radically in their procedures of verification. How they convince and what they convince us of are also different. That is, well-formed arguments convince us of their truth, whereas stories convince us of their lifelikeness. One verifies by eventual appeal to procedures for establishing formal and empirical truth, whereas the other does not establish truth but verisimilitude.

According to Schwandt, in discussions about the methodologies of qualitative inquiry, the term *verisimilitude* is used in three ways, all dealing with the quality of the text:

1. Verisimilitude as a criterion (others include plausibility, internal coherence, and correspondence to readers' own experience) sometimes cited as important for judging narrative inquiry. A narrative account (referring either to the narratives generated from or by respondents or to the narrative report produced by the inquirer) is said to exhibit the quality of verisimilitude when it has the appearance of truth or reality.

2. Verisimilitude as a criterion for judging the evocative power or sense of authenticity of a textual portrayal: A style of writing that draws readers into the experiences of respondents in such a way that those experiences can be felt.

3. Verisimilitude as the relationship of a particular text to some agreed on opinions or standards of a particular interpretive community. A particular text (e.g., book review, scholarly essay, speech to a scholarly society, research report, and so on) has verisimilitude to the extent that it conforms to the conventions of its genre.

Schwandt, 1997, pp. 170-171

With verisimilitude rather than validity to the fore, the views of Bruner (1986) are reinforced by those of Eisner (1991), who argues that what is personal, literary, and even poetic can be a valid source of knowledge, even though this is consistently denied by those who hold a restricted "scientific" view of what constitutes knowledge and truth. Eisner acknowledges that works of poetry and literature are not true in the *literal* sense, but suggests that they can be true in the *metaphorical* sense. To restrict truth to literal truth alone is to restrict knowledge to those forms of discourse that can be literally true.

> Scientific knowledge is seldom true in the literal sense. . . . Especially in the social sciences where metaphor, analogical reasoning, and hypothetical constructs abound, literal truths are scarce. When we use literature, for example, to enlarge understanding, *literal* truth becomes an irrelevant criterion for appraising its utility. A piece of fiction can be true and still be fiction. Fiction, in the metaphorical sense, is "true to life"; it helps us to perceive, experience, and understand what we have previously neglected.
>
> *Eisner, 1991, p. 108*

More recently, Eisner (2001) recalls an interview with an important American writer who stated that "To be great fiction it has to be true." This comment led Eisner to reflect on how, in making a distinction between fact and fiction, he was on slippery ice.

Yet, while there is a sense in which great fiction is true—the way in which the work of Chekov is true, or Mark Twain, or Charlotte Bronte, or Wallace Stegner—true in the way it displays the universal in the particular, there is a sense in which fiction is not true. It's not true in the literal sense. Can fiction count as a research genre? I believe it can *if* it is true in the former sense. Whether people will regard it as such will depend on the quality of their education.

Eisner, 2001, p. 140

Sandelowski (1994) supports this view when she argues that artistic truths are often more true to life than scientific ones because they are able to provide us with visions of human nature more resonant with our own experiences than any psychological, sociological, or other conventionally scientific rendering of it. For her, "What differentiates the arts from science is not the search for truth per se, but rather the kinds of truth that are sought" (p. 52). In this sense, different tales can seek to explore different truths about the same phenomenon. Thus, we can learn about the same thing from many different approaches and sources. In this context, diversity needs to be seen as a strength, not a weakness.

Starting Points

According to Tierney (1999), if the old frameworks used for passing judgment on research are shopworn and no longer of use to many of us, and if new structures are useful, the next step is to develop ways to think about, judge, and improve the variety of genres that we use. Therefore, as new representational practices emerge, then new standards for judging the postmodern text need to be developed. One simple rulebook, Tierney argues, will no longer suffice, and one of the immediate challenges is to come to terms with criteria for goodness. So how might this be done?

One way might be to reflect on the "emerging criteria" (Lincoln, 1995) that are being suggested by leading scholars for passing judgment on different forms of qualitative research and different kinds of representation. Accordingly, my tactic in the following sections is to offer a number of criteria lists that have been put forward in recent years. For example, in relation to what makes a "good" narrative study, Lieblich et al. suggest four criteria:

1. *Width: The comprehensiveness of evidence.* This dimension refers to the quality of the interview or observations as well as to the proposed interpretation or analysis. Numerous quotations in reporting narrative studies, as well as suggestions of alternative

explanations, should be provided for the reader's judgment of the evidence and its interpretation.

2. *Coherence: The way different parts of the interpretation create a complete and meaningful picture.* Coherence can be evaluated both internally, in terms of how the parts fit together, and externally, namely, against existing theories and previous research.

3. *Insightfulness: The sense of innovation and originality in the presentation of the story and its analysis.* Close to this criterion is the question of whether reading the analysis of the life history of an "other" has resulted in greater comprehension and insight regarding the reader's own life.

4. *Parsimony: The ability to provide an analysis based on a small number of concepts, and elegance or aesthetic appeal* (which relate to the literary merits of written or oral presentations of the story and its analysis).

Lieblich et al., 1998, p. 173

Lincoln and Guba put forward the following *authenticity* criteria for judging the processes and products of naturalistic or constructivist inquiry (see also Manning, 1997):

1. *Fairness* was thought to be a quality of balance; that is, all stakeholder views, perspectives, claims, concerns, and voices should be apparent in the text.

2. *Ontological and educative authenticity* were designated as criteria for determining a raised level of awareness, in the first instance, by individual research participants and, in the second, by individuals about those who surround them or with whom they come into contact for some social or organizational purpose.

3. *Catalytic and tactical authenticity* refers to the ability of a given inquiry to prompt, first, action on the part of the research participants, and second, the involvement of the researcher/evaluator in training participants in specific forms of social and political training if participants desire such training.

Lincoln & Guba, 2000, pp. 180-181

L. Richardson offers five criteria she uses when reviewing papers or monographs submitted for social scientific publication.

1. *Substantive contribution:* Does this piece contribute to our *understanding* of social life? Does the writer demonstrate a deeply grounded (if embedded) social scientific perspective? How has this perspective informed the construction of the text?

2. *Aesthetic merit:* Does this piece succeed aesthetically? Does the use of creative analytical practices open up the text, invite interpretive response? Is the text artistically shaped, satisfying, complex, and not boring?

3. *Reflexivity:* Is the author cognizant of the epistemology of postmodernism? How did the author come to write this text? How was the information gathered? Are there ethical issues? How has the author's subjectivity been both a producer and a product of this text? Is there adequate self-awareness and self-exposure for the reader to make judgments about the point of view? Does the author hold him- or her-self accountable to the standards of knowing and telling of the people he or she has studied?

4. *Impact:* Does this affect me? Emotionally? Intellectually? Does it generate new questions? Move me to write? Move me to try new research practices? Move me to action?

5. *Expression of a reality:* Does this text embody a fleshed out, embodied sense of lived experience? Does it seem "true"—a credible account of a cultural, social, individual, or communal sense of the "real"?

L. Richardson, 2000, p. 937

In telling an ethnographic short story about how she *creates* criteria when evaluating alternative genres of writing, Ellis (2000) notes that on the first reading the work needs to engage her so that she reads the whole paper without stopping to evaluate cognitively from a distance. It also has to evoke a response, making her reflect on details from her own experience, memories, and feelings. If the story has evoked Ellis narratively, then she asks a series of questions to explore the contribution more fully and more cognitively. These questions include the following:

I ask what I have learned from the story: About social life, social process, the experience of others, the author's experience, my own life. Is there anything "new" here or a new way to view or twist the familiar?

I ask about the plot of the story: Does the story have a balance of flow and authenticity of experience? Has the author been able to represent the chaos, yet do it in a way that provides a readable and understandable experience? Is there sufficient, yet not overblown dramatic tension? Do I long to turn the pages to find out what happened? Are there unexplainable holes in the plot? Or too much detail about insignificant points? Is the story coherent and logically consistent? Is there a sense of verisimilitude? Of course, I don't think of that word—instead I ask, does the story ring true, is it lifelike?

I ask about the writing of the story: Does the author show instead of tell? Does she develop characters and scenes fully? Are there too many characters and scenes to follow? Does she edit so that all words are necessary, well placed, and the best choices? Does she paint vivid pictures? Sounds? Smells? Feelings? Does the conversation feel real to life? Did the author know the end of the story when she started or does writing become a form of inquiry? Is the story sufficiently complexified and nuanced? Is there a literary sensitivity to the writing? Does the ending surprise or move me, making me think about the story in a new manner or see connections or the whole in a way I had not seen before? Are there so many spelling and punctuation mistakes that they interrupt my reading?

I ask about the goals, claims, and achievements of the author: What is the author trying to achieve? Has he achieved his goals? Are they worthwhile goals? Are these goals that can be met by this writing form? Might there be a better way of achieving his purposes? Can the author legitimately make these claims for his story? Did the author learn something new about himself? About other characters in the story? About the processes and relationships described? What might readers take from the story? Will this story help others cope with or better understand their worlds? Is it useful?, and if so, for whom? Does it encourage compassion for the characters? If not for the characters, does it encourage compassion for those acted upon by them? Does the story promote dialogue? Does it have the potential to stimulate social action?

I ask about ethical considerations: Did the author get permission to portray others? Give them a chance to contribute their perspectives to the story? If not, are there sufficient and justifiable reasons why not? Are other characters sufficiently complexified? Is the author? Is this exclusively the author's interpretation of what is going on? Does the contribution of the story outweigh conceivable ethical dilemmas and pain for characters and readers?

Ellis, 2000, pp. 275-276 (my italics)

Finally, Pelias (1999), focusing specifically on the *poetic essay* as a mode for rendering performance, asks, What might constitute an acceptable and authoritative account? He notes that the poetic essay seeks a different standard for presenting the performance on the page and offers four criteria: coherence, plausibility, imagination, and empathy.

A coherent poetic essay holds together, gels in an intelligible and articulate manner. Its parts seem to coalesce, to become intertwined, to find relationships with one another. The parts may settle into a seeming unity or may shatter into a disjunctive array. In either case, the parts insist upon some association that yokes them together. As the parts come together in their harmonious and inharmonious combinations, the essay finds its voice, a voice that often cannot be contained within a single speaker.

The plausible poetic essay appears credible. It pulls together a believable combination of the parts. Like a good story, it offers a convincing narrative. It stands as a version, an interpretation among many that appears reasonable to accept. It seeks an internal logic, one that may be filled with ambiguity, tension, and contradictions. Held against the external world, it may echo or challenge everyday understandings. Its account, then, is a temporary diagnosis. It illustrates the possible.

The imaginative poetic essay is literary. It calls upon traditional aesthetic standards, those questioned by literary critics and relied upon by creative writers. It privileges the sensuous, the figurative, the expressive. It calls for an aesthetic transaction, an encounter between the writer and the reader. It demands engagement. Like good phenomenology, it presents through reflection and imaginative free variation the complexity and richness of its subject.

The empathetic poetic essay is marked by respect. It strives to feel with others, to understand what others see. It works for a generosity of spirit that creates space for others. It invites dialogue. It is an open invitation for speech, a desire to hear others. The empathetic essay, then, privileges an ethics of fairness, sensitive to the ideological consequences of its own discourse and aware that an empathetic gesture cannot become a substitute for political action.

Pelias, 1999, p. xiii (my italics)

These lists are starting points from which the selection of criteria for judging different kinds of tales can begin. For example, different combinations might be chosen for judging an autoethnography as opposed to an ethnographic fiction. Indeed, given that the criteria chosen will depend on the context of the reading as well as the purpose of the telling, the criteria might well change over time. The temptation to make definitive statements as to what makes, for example, a "good" ethnodrama across all contexts and conditions and for all purposes should be avoided. Furthermore, none of these lists should be taken as complete in itself. These lists are not closed; they can be added to and subtracted from as the form and purposes of inquiries change.

Selecting From Lists

It is interesting to note how various authors have made suggestions for selecting criteria they see as appropriate for judging specific kinds of tale. For example, Van Maanen (1988) emphasizes that the standards for impressionist tales are not disciplinary but literary. Artistic nerve is required of the teller, and literary standards are of more interest than scientific ones.

> The audience cannot be concerned with the story's correctness, since they were not there and cannot know if it is correct. The standards are largely those of interest (does it attract?), coherence (does it hang together?), and fidelity (does it seem true?) . . . the main obligation of the impressionist is to keep the audience alert and interested.
>
> *Van Maanen, 1988, p. 119*

Similarly, in reflecting on how his own ethnographic fiction might be judged, Tierney (1993) points out that if his story "works" and serves its purpose then it does so from a *literary* perspective. For him, such a perspective reframes our analysis away from scientific standards of validity or trustworthiness and toward more literary definitions of good literature.

> The reader does not judge the text according to standardized scientific criteria, or with the assumption that the text is meant to explain all such situations. Rather, the reader judges the text in a self-referential manner and might call upon the following questions in judgment: (a) Are the characters believable? (b) Are there lessons to be learned from the text for my own life? (c) Is the situation plausible? (d) Where does the author fit in the formation of the text? (e) What other interpretations exist? and (f) Has the text enabled me to reflect on my own life and work?
>
> *Tierney, 1993, p. 313*

A. Frank, in writing about his own experiences as a cancer survivor in *At the Will of the Body*, offers the following hope:

> I want what I have written to be touched as one touches letters, folding and refolding them, responding to them. I hope ill persons will talk back to what I have written. Talking back is how we find our own experiences in a story someone else has written. The story I tell is my own, but readers can add their own lives to mine and change what I have written to fit their own situations. These changes can become a conversation between us. . . . My own experiences are in no sense a recipe for what others can expect or should experience. I

know of no exemplary way to be ill. We all have to find our own way, but we do not necessarily have to be alone.

Frank, 1991, pp. 4-5

With regard to her own autoethnography about a nine-year relationship with her partner, who died from emphysema, Ellis (1995a) highlights the issue of evocation. She argues that as a form of knowing, the "validity" of evocative storytelling is best judged by whether it evokes in the reader a feeling that the experience described is authentic, believable, and possible. For her, the "generalizability" of the story is best judged by whether it speaks to readers about their own experiences. Ellis asks the following questions:

Did my story engender conversational response toward the text as you read? Did the story illustrate particular patterns and connections between events? Did you want to give the story to others to read because you think it speaks to their situation? How useful would this story be as a guide if you encountered a similar experience in your life? . . . What text did you, the reader, create of my story? Did this narrative make you think about or shed light on events in your own life? Would you have acted differently than we did? Would you have told this story the way I told it? Did the words I wrote elicit from you an emotional response to examine? What did you learn about yourself and your relationships through your responses to my text?

Ellis, 1995a, p. 319

Four of the six reviewers of my own autoethnography, "The Fatal Flaw" (Sparkes, 1996), described it also as a provocative, engaging account, with the story grabbing the readers, making them think and feel, and bringing them into their own bodies (see Sparkes, 2000). I am also reminded of the reactions of many of the undergraduate students at Exeter University who, as part of their qualitative research module in their second year, read this article. For example, Lax (1999) opted to construct an autoethnography / narrative of self about his own sporting injuries for his dissertation topic. In providing a rationale for his choice of approach, he notes how "The Fatal Flaw" held his attention more than any other journal article had.

There were three main reasons for this; the difference in style to the usual journal fare; the raw honesty of the piece. Here was one of my lecturers opening up his private life to the class (not just the invisible audience of the readership of the journal). I doubt I would show this piece [if I had written it] to many people on my course; finally, but mainly, because it struck so many chords with my own life, my own story. I made marks in the margins where something resonated with my own story, my own life. I showed it to my twin brother. I showed it to my friends. It left a mark, which I believe is one of the reasons it was written.

Lax, 1999, p. 20

Evoking a response from readers requires that they experience vicariously the life of another, even if only momentarily. To assist this process, it helps if the tale is deemed to be *authentic.* On this issue, Lincoln (1993) argues that authenticity emerges when a text not only is faithful to the storylines of the teller but also conveys the *feeling tone* of life as lived. This feeling tone is best conveyed when the text itself invites the reader into a vicarious experience (however brief) of the life or lives being described. If this invitation is taken up, then the reader might gain an experience of the lives "in the round," with a range of mood, feeling, experience, situational variety, and language. Consequently, the reader can come away from the text with a heightened sensitivity to the life or lives being depicted, and with some flavor of the kinds of events, characters, and social circumstances that circumscribe those lives.

Here, it is interesting to note the reactions of two people to my ethnographic fiction about Alexander, a gay sportsman and physical education teacher (Sparkes, 1997a). The first reaction is from a gay friend who was not based in physical education but who worked in the education field and was aware of the dynamics of the physical education culture. These are his comments:

I'm not really sure what sort of response you "want." The first thing that strikes me is the *authenticity* of Alex's voice—this is a thoroughly plausible story—to me at least. The relationship with the father doesn't ring as true, though. Most gay men have difficult paternal relationships—often completely rejecting/rejected. I don't quite see how Alex's love for his father transcends his father's bigotry—not something I can resonate with at all, or find authentic from my wide experience of (usually very out, rather camp, and non-Jock admittedly) gay men. Knowing the jock-culture of Exeter, I would imagine that the authenticity (overall) of the story would profoundly disturb some of the respondents who'd find it rather "too near the bone"—especially since Alex is in charge of his own destiny and eschews the jock culture to be fully his "own man." The very fact of this "happy/satisfying" resolution is almost calculated to raise the ire of the self-repressed gay in the jock culture. I'd be interested to know what the responses from *within* the jock culture were. I enjoyed reading this and found it reassuring!

The second reaction is from a gay, male physical education student at another institution, who gained access to Alexander's story via one of his lecturers and contacted me. The following are some extracts from his response to the story:

> I found Alexander's story to be very authentic. Although his story is slightly different from my own, I still found it very believable. I come from a middle class family who are very worldly. I had no trouble telling my parents that I was gay, although it has taken them a long time to believe me. I think they thought being very sporty and masculine did not go with being gay. . . . Most of my sporting friends at my tennis and squash club knew I was gay during my early twenties. Football teammates did not know, except two, but did not take me seriously. . . . I am pleased you are giving homosexuals a voice. . . . Even though you are not a homosexual I still believe your voice is important. . . . Any voice on the subject is good to hear. . . . But a believable story, a lot of truth in it.

Issues of authenticity are connected to what is *believable* and possible. According to Rinehart, "For ethnographic writers who use fictional methods in their writing, representing well and believably is crucial" (1998a, p. 205). For him, uncommunicative writers in the social sciences become so immersed in the "facts" that they fail to realize that they have not engaged the readers. They do not produce fictions that have a visceral *feel*.

> Believability is dependent on the kind of description that is accurate in a holistic, evocative, emotionally engaging sense. And that description relies on glimpses of telling detail more than on total immersion in detail. What is enough detail, and what is too much detail is one of the fundamental differences between believable and cloying, between engaging and pedantic, between creatively accurate and merely replicable and boring writing.
>
> *Rinehart, 1998a, p. 205*

Such issues are important because, as Ceglowski (1997) reminds us, "researchers must write vital texts because qualitative research relies solely on narrative text to convey meaning. If the researcher cannot engage the reader in the text, then it is less likely that the reader will finish the text" (p. 193). Here, I am reminded of the responses of the students I work with to my ethnographic fiction (Sparkes, 1997a). Once the story has been handed out and students begin to read, the room invariably goes silent. Nobody speaks, even when they have finished. As I watch, I get a real sense of the students being taken into the story. Once all have finished, I ask, "What do you make of the story?" This simple question opens the floodgates to what has been with all groups an energetic, stimulating, insightful, and at times emotionally tense debate. Their reactions suggest to me that the character of Alexander is certainly believable, and that the situation he confronts is plausible to the students.

They certainly learn lessons about their own lives, particularly in relation to their acceptance and tolerance of homophobia. Likewise, it is clear from the comments made both in the lecture and on their written sheets that this fiction stimulates the students to reflect on their own lives and work. It would seem that the story I constructed about Alexander manages to cut through the scientized comfort zone of many of the students and allows them to explore often ignored or repressed dimensions of their own subjectivity.

Of particular interest here is the reaction from the students when, toward the end of the session, I inform them that I have written the story—that I have "invented" Alexander, that he is "fictional" and not "real." I explain that although many parts of the story draw on my own "straight" sporting experiences and conversations with gay friends as material, the tale is fictional. Many students simply do not believe me and think I am trying to protect my "source." Others are prepared to accept that I invented Alexander but are adamant that he is "real" in the sense that he is "believable" and that this is a "true" portrayal of what it might be like to be gay and to live in the sport and physical education culture. That is, Alexander is deemed to be "true to life." Interestingly, in our ongoing discussions, even when they knew the story was a fiction, the students continually referred to Alexander *as if* he were a real person. In doing so they continued to read themselves into his storyline and reflect on his world in relation to their own.

The criterion of believability is closely linked to that of *fidelity*. The latter is a criterion for practicing and evaluating narrative inquiry that links social science and art. For Blumenfeld-Jones (1995), truth is *what happened in a situation* (the truth of the matter), and fidelity is *what it means to the teller of the tale* (fidelity to what happened to that person). Drawing on work in aesthetic philosophy, Blumenfeld-Jones also establishes the criterion of *believability* in relation to fidelity. A narrative is believable when it can be credited with conveying, convincingly, that the events occurred and were felt in ways the narrator asserts. It is also believable when it resonates with the experiences of the audience.

To assign believability, audiences must experience a congruence with their own experiences of similar, parallel, or analogous situations. They do not have to derive the same meanings as the artist's original meaning. This is what provides art (and narrative inquiry) with its power of redescription of reality. Beyond such believability, the audience must be able to find that the artist's (or narrative inquirer's) version of reality can usefully or meaningfully redescribe their situations. The audience members may think that the artist (narrative inquirer) is out of touch with reality or simply cannot see other contingencies in the situation. Believability and fidelity ("This situation seems to have happened") need not be damaged by such a judgment ("I would have made a different interpretation of it").

Blumenfeld-Jones, 1995, p. 31

In calling on criteria like verisimilitude, coherence, evocation, empathy, authenticity, fidelity, and believability, a move begins toward incorporating more literary and artistic forms of judgment in the social sciences. In saying this, I do not wish to romanticize the world of literature. For example, it would be simplistic to assert that if an autoethnography, ethnographic fiction, or poetic representation were judged by a number of novelists or essayists, there would be consistency and agreement on the criteria used. Tensions and contradictions would likely remain. Likewise, in terms of their positioning and making passing of judgment, literary reviewers need be no less dogmatic and inflexible than reviewers in the social sciences. At best we might suggest that the former could, and would, have called on different criteria to judge and respond to such tales. Thus, different criteria are available, and they can be used.

Continuing Dilemmas: Is This Real Research?

Transgressing the boundaries of traditional forms remains problematic in terms of judgment and status in the community of scholars. For example, Ceglowski notes, "by blurring the boundaries with literary devices such as short stories, researchers may jeopardize the scientific standing of their field" (1997, p. 194).

As an example of this dilemma, I once again turn to reactions to my own autoethnography, "The Fatal Flaw" (Sparkes, 1996). The published version of this article continues to raise questions. One example will have to suffice. In describing "The Fatal Flaw," Coffey notes that I take a different approach to the writing of the self and that my "ethnographic body narrative is extremely personal and highly reflective" (1999, p. 124). Coffey argues that my body and self are not adjunct to the writing, but centrally positioned, which gives a performative quality to my "autobiographical ethnographic project" (p. 124). She then suggests that I present a rather different, self-centralized approach to ethnographic writing and tales of the body in which my combination of the personal and the ethnographic is explicit and foreshadowed, so as to form the pivot rather than the apology in the text.

> The account can be located within a new wave of autobiographical writing which focuses on the body and the self. They are qualitatively different from the confessional or semi-autobiographical accounts. . . . These kinds of tales use ethnographic analyses and writing to explore the self. They centralize rather than add the self. The emphasis is on the analysis and the writing of the self, and the relationship between self and the field. These texts do not apologize for the presence of the self, nor are they simply accounts of the research process. . . . Rather the self and the field are seen as symbiotic, and the writing seen as establishing the interconnectedness of the two.
>
> *Coffey, 1999, p. 125*

Coffey (1999) states that the move toward more autobiographical, ethnographic texts has been encouraged by the literary turn in the social sciences generally, while other contemporary movements, such as postmodernism and feminism, have encouraged texts that are more emotional, personal, and complex. She also notes that "There is, of course, a debate as to whether these personalized accounts actually constitute ethnographic writing" (p. 125). For her, the boundary between ethnography and autoethnography has become difficult to navigate.

Again, using "The Fatal Flaw" as an exemplar, Coffey (1999) first acknowledges that I create a powerful and compelling narrative of the self and the body, and that by using ethnographic writing as analysis, autobiography, diary, and self-discovery, I employ different textual formats to create a multilayered narrative of my body, family, self, and experiences. Then, Coffey suggests that what I am actually engaged in is the production of autoethnography or ethnoautobiography. Here, following Bochner and Ellis (1996a, 1996b), ethnographic writing is used as a form of creative nonfiction, where data are connected to experiences and to an audience.

Given that this form of ethnographic practice is as much about writing *to* as it is about writing *about*, Coffey (1999) is left with some problems. For example, where does ethnography stop and autobiography begin, given that writing the self and the writing to significant others of the self can be classed as both? Perhaps, Coffey suggests, the introduction of the new literary forms described means we are now standing on the boundary between ethnography and autobiography. What we might be witnessing, in papers like "The Fatal Flaw," could be a new form of ethnographic practice, more firmly rooted in a social context and the situatedness of authorself.

This may have positive consequences for the representation of peopled, polyvocal social worlds. Yet some would say that such texts are not "doing" ethnography at all, but are self-indulgent writings published under the guise of social research and ethnography. Rather than utilizing literary and autobiographical devices to write ethnography we may be witnessing the use of ethnographic devices to write autobiography. . . . It remains debatable as to whether utilizing ethnographic strategies to write autobiography really "counts" as ethnography at all.

Coffey, 1999, pp. 155-156

Of course, *what counts* brings us back full circle to the criteria used to judge what counts in the first place.

 ## Reflections

Given that we are able to consider various combinations of criteria for judging different tales, it is clear that not "anything goes" in new writing practices. Indeed, as L. Richardson points out, "I believe in holding CAP [Creative Analytical Practices] ethnography to high and difficult standards; mere novelty does not suffice" (2000, p. 937). Judgments are made, and must be made. Therefore, not all ethnographic fictions, creative fictions, ethnodramas, poetic representations, autoethnographies, or confessional tales will be defined as "good" by the research community. Claims, or fears, to the contrary are usually based on a misguided notion of relativism and its consequences for social inquiry. These misguided views have been criticized elsewhere (see Bernstein, 1983; Eisner, 1991; Feyerabend, 1987; Geertz, 1984; K. Gergen, 1999; Rorty, 1985; J. Smith, 1988, 1993; J. Smith & Deemer, 2000; Sparkes, 1991). Several points, however, are worth mentioning. First, as Rorty points out, at least three different views are associated with relativism:

The first view is that every belief is as good as every other. The second is the view that "true" is an equivocal term, having as many meanings as there are procedures of justification. The third is the view that there is nothing to be said about either truth or rationality apart from descriptions of the familiar procedures of justification which a given society—*ours*—uses in one or another area of inquiry.

Rorty, 1985, pp. 5-6

The first view is a foolish position to adopt and is rightly criticized. For the most part, given their starting assumptions, qualitative researchers hold to the third view and fail to see what all the fuss surrounding relativism is about. First, as Geertz (1984) notes, the fear of relativism is unfounded because the moral and intellectual consequences that are supposed to flow from it—such as subjectivism, nihilism, incoherence, Machiavellianism, ethical idiocy, aesthetic blindness, and so on—do not in fact do so. Equally, the promised rewards of "escaping" relativism, which tend to revolve around the fond hope of developing a pasteurized form of knowledge, are illusory. As Geertz concludes, relativism today serves largely "as a specter to scare us away from certain ways of thinking and towards others" (p. 263). Furthermore, as J. Smith (1988) argues, the fears so often expressed in relation to relativism have force *only* if it is assumed that social inquirers can objectively ground their knowledge claims. Once, he suggests, it is realized that objectivism is impossible, relativism can be seen for what it actually is, "the inevitable consequence of our hermeneutical or interpretive mode of being in the world" (p. 18).

As such, relativism is not something that can be transcended but something that we must, as finite beings, learn to live with. On this issue J. Smith (1993) suggests that all an understanding of our human finitude brings us is the realization that there can be no fixed standard, historical or contextual, on which to base our judgments. Therefore, just as with our inquiries we construct reality as we go along, with these inquiries we also construct our criteria for judging as we go along. Accordingly, in relation to judging qualitative research in general, and different tales in particular, it is possible to advocate a view that is both relativistic and pluralistic, but not mindless. That is, qualitative researchers can and do make judgments, and they can and do resolve disagreements between themselves over research findings and various ways in which these findings can be represented.

Unfortunately, work like confessional tales, autoethnography, ethnodrama, poetic representations, and fictional representations, which operate on the borderlines of disciplines and cross or blur boundaries, seem to cause problems for those obsessed with *criteriology*, or the constant search for permanent or universal criteria for judging research. Garratt and Hodkinson (1998) have commented how, for some, this has created a desire to remove the uncertainties associated with the twin crises of representation and legitimation. They point to the alternative offered by Schwandt (1996), who proposes that social inquiry is redefined through the application of practical philosophy. This redefinition involves challenging the ideology of "epistemic criteria," with its parallel focus on fixed and predetermined rules, and invoking a new moral and political framework within which values and concerns may be addressed through open dialogue, critical reflection, and a willingness to change. In this respect,

Schwandt talks of criteria as *enabling conditions* that not only allow for the development of *guiding ideals* that can be used to facilitate the adjudication of research but also actively engage researchers in a process of deliberation that is characterized by coherence, expansiveness, interpretive insight, relevance, rhetorical force, beauty, and texture of argument.

Such views echo the thoughts of J. Smith (1989, 1993), who proposes that judgments in qualitative inquiry take place through debate, discussion, and the use of exemplars. Here, qualitative researchers do not appeal to a supposedly external referent point or set of "facts" that exist independently of themselves and their historical conditions, which can be known through specified procedures. Instead, Smith (1993) points out, disagreements are resolved because researchers come to share similar interests, purposes, and values. Therefore, the task of reaching agreement and passing judgment becomes a practical and moral one rather than an epistemological one. This involves "a willingness to engage in a free and open exchange of reasoned arguments over why one researcher's interpretation is more appropriate than another researcher's interpretation" (p. 119).

As part of this exchange of reasoned argument, researchers will certainly draw on various criteria. However, as J. Smith (1993) notes, the term "criteria" can have different meanings. It can mean a *standard against which to make judgment.* Here, the term is laden with foundational implications. That is, it becomes a touchstone that can be employed to distinguish the good from the bad, the correct from the incorrect. In contrast, Smith points out, "criteria" may refer to *characterizing traits* that have, at best, mild implications as a prescription for inquirer behavior and do not necessarily refer to something that is held to be foundational. Here, researchers might discuss the characterizing traits of a particular approach to inquiry in sport and physical activity and simply note that these criteria are the way researchers seem to be conducting their particular kinds of inquiries at the moment. The difference here from the foundational view is that qualitative researchers are willing to describe what one *might* do, but they are not prepared to mandate what one *must* do across all contexts and on all occasions.

For J. Smith (1993) and J. Smith and Deemer (2000), once criteria come to be seen as characterizing traits or values that influence our judgments, then any particular traits or values will be subject to constant reinterpretation as times and conditions change. A characteristic of research we thought important at one time and in one place may take on diminished importance at another time and place. Therefore, various criteria in list form may act as a starting point for judging a certain kind of tale, but these may not apply on all occasions; other criteria can be added to or subtracted from them depending on the circumstances. These lists are challenged, changed, and modified in their application to actual inquiries and actual writing practices. As Smith and Deemer emphasize, the limits of modification are a *practical* matter; they are worked and reworked in the context of actual practices and applications and cannot be set down in abstract formulas. "Our lists are challenged, changed, and modified not though abstracted discussions of the lists and items in and of themselves, but in application to actual inquiries" (p. 889).

The flexibility of nonfoundational lists is important to emphasize lest one form of dogma be replaced by another in the face of the chronic uncertainty that we now have to live with as part of the postmodern condition. As Garratt and Hodkinson (1998) are well aware, if different groups applied different but equally rigid predetermined research criteria then the end result would be the Balkanization of the research community. In making a powerful case for the hermeneutics of interpretation and understanding research as *experience*, they argue against choosing any list of universal criteria in advance of reading a piece of research. To do so would foist on research artificial categories of judgment, preconceptions of what research should be, and a framework of a priori conditions that may be impossible or inappropriate to meet, at least in some cases. For Garratt and Hodkinson, "The selection of appropriate preordained sets of different paradigmatic rules, then, is not a solution. A more constructive way forward begins with the acknowledgment that the selection of criteria should be related to the nature of the particular piece of research that is being evaluated" (p. 527). Or, as J. Smith (1993) would have it, we need to construct our criteria for judging various forms of inquiry as we go along.

Acknowledging the flexibility of nonfoundational lists is particularly important in guarding against the Balkanization of new writing practices and forms of representation in sport and physical activity. Here, there would only be one way to write a good autoethnography, only one way to write a good ethnographic fiction, and so on. In her review of a number of books that advocate experimental writing as ethnographic alternatives, Rath (2001) explains her concerns:

My main concern is that these self-labelled "alternative writing forms" are at risk of solidifying as they gain acceptance to a new, albeit minority, academic canon. Academic precedence is a great comfort to authors. Contextualizing a writing style via an alternative canon generated by established academics and judged by accepted criteria, has to be a seductive option for authors attempting to transgress traditional text boundaries. Therefore, while I applaud the efforts made by these three books, I am compelled to sound a warning shot—how long will it be before the forms admitted by this groundbreaking series will become the only way to script an "ethnographic alternative"?

Rath, 2001, p. 114

As Schwandt (1996) points out, criteria need to be seen as enabling conditions to be applied contextually. Then the terrain on which judgments get made about different forms of qualitative research and different genres of representation will be continually changing and will be characterized by openness rather than stability and closure. Here, as part of what I have described as a polyvocal research community (Sparkes, 1991), plurality of judgment would be seen as a strength and a sign of intellectual vibrancy (see also Sparkes, 1989, 1992, 1998c, 2001). Various criteria from various traditions might be called on in combination to judge the quality of a piece of research. As DeVault (1997) notes, as autoethnographies and narratives of self become more common among social scientists, researchers will need to develop new avenues of criticism and praise for such work: "One element in this new evaluative understanding might be a clearer sense of how to combine 'scientific' with 'literary' standards, without mystifying the latter" (p. 224).

Equally, new criteria might need to be developed for certain contexts and purposes. For example, Blumenfeld-Jones (1995) talks about the *emerging* criteria, not set or firm criteria, in relation to judging narrative studies. For example, with regard to fidelity and believability, he suggests that there may be much variation in how these criteria are understood and this could manifest itself in the way a reviewer of a piece of narrative inquiry communicates to the inquirer:

> Perhaps he or she would be forced to define these terms freshly and speak to the inquirer in terms of rationales and explicit criteria. This would encourage an evaluation situation in which both the reviewer and inquirer would develop new understandings and not merely stand in judgment over one another. . . . if narrative inquiry is a type of hermeneutic act, then the criteria which we apply to it also ought to be hermeneutic in character.

Blumenfeld-Jones, 1995, p. 33

This point is reinforced by J. Smith and Deemer (2000), who emphasize that in any encounter with a production, especially something "new" (e.g., an ethnodrama or an ethnographic fiction about sport and physical activity), one must be willing to risk one's prejudices: "Just as in the process of judgment one asks questions of a text or person, the person or the text must be allowed to ask questions in return" (p. 889). They argue that

223

approaching a novel piece of work requires that one be willing to allow the text to challenge one's prejudices and possibly change the criteria one is using to judge the piece, thereby changing one's idea of what is and is not good inquiry. Smith and Deemer are quick to point out that to be open does not mean to accept automatically, and that one may still offer reasons for not accepting something new. They also emphasize that there is no method for engaging in the risking of prejudices: "If anything, to risk one's prejudices is a matter of disposition—or, better said, moral obligation—that requires one to accept that if one wishes to persuade others, one must be equally open to being persuaded" (p. 889).

Being open to persuasion and risking prejudices in a world of multiple perspectives means that there will undoubtedly be tensions, contradictions, conflicts, and differences regarding the criteria deemed appropriate for judging the meaning and quality of different forms of inquiry and different forms of representation. This should not cause undue anxiety. Rather, as Garratt and Hodkinson (1998) emphasize, such diversity should be seen as an invitation to deepen our understanding and sharpen our judgments of those specific pieces, of the issues they raise, and of research in general. If confessional tales, autoethnography, ethnodrama, poetic representations, and fictional representations in the field of sport and physical activity do nothing else but stimulate us to think about such issues, then they will have made a significant contribution.

Taking an optimistic view, we need to recognize that standards do change. For example, many once believed that the passive voice was the only way to write a dissertation, but such a belief is now rejected in a multitude of areas. Not so long ago, qualitative research in general was fighting for acceptance, but it is now found in a host of prestigious journals and is taught in classes at numerous institutions of higher education. The use of fictional representations would have been unthinkable not so long ago. Yet, recently Eisner asked, "who would have predicted a decade ago that fiction might be considered a legitimate form for a Ph.D. dissertation" (1997, p. 263). Therefore, as things change so will the stories we tell one another, along with the criteria we use for reading stories. Indeed, as L. Richardson asserts, "it is our continuing task to create new criteria for choosing criteria" (1999, p. 665). All of which presents researchers in sport and physical activity with numerous exciting challenges in the years ahead. There is much to look forward to.

Epilogue

If I had tried to write this book 10 years ago, I would have struggled. Certainly, I could have provided examples of scientific and realist tales from the world of sport and physical activity, but I would not have been able to find examples of confessional tales, autoethnography, poetic representations, ethnodrama, or ethnographic fictions. Even now, the literature in sport and physical activity is not awash with these alternative tales. The fact that there are examples at all, however, and that scholars in these domains are beginning to experiment with new forms of representation and writing practices is, for me, significant—and encouraging.

The growth of new writing practices in sport and physical activity signals a growing awareness in this community that there is no single way, much less one "right" way, to stage a text or to know about a phenomenon. As L. Richardson comments:

> How we write has consequences for ourselves, our disciplines, and the publics we serve. *How* we are expected to write affects *what* we can write about; the form in which we write shapes the content. Prose is the form in which social researchers are expected to represent interview material. Prose, however, is simply a literary technique, a convention, and not the sole legitimate carrier of knowledge.
>
> *Richardson, 2001b, p. 877*

Following L. Richardson, we are beginning to realize that, "Like wet clay, the material can be shaped. Learning alternative ways of writing increases our repertoires, increases the numbers and kinds of audiences we might reach" (2000, p. 936). This experimentation also signals a growing sophistication in the research community regarding the nature of qualitative inquiry and what it can contribute to our understanding. In this sense, various scholars in sport and physical activity are taking the next step in the development of a qualitative research informed by the new philosophy of science, which, according to Polkinghorne, "is to move out from under the conventional format for reporting research" (1997, p. 18). Finally, it indicates the growing confidence of scholars to grapple with some of the extremely difficult and profound dilemmas that go hand in hand with the qualitative venture as we enter the 21st century.

How we present our work, and to whom, now are more up for grabs than ever before. For me, as for Richardson, "Alternative representational methods create a welcoming space in which to build a community of diverse, socially engaged researchers in which everyone profits" (L. Richardson, 2001b, p. 888). As Anderson comments, this makes for exciting times as the new fields and new writing genres that have opened to ethnography "offer those of us who want to stretch our wings, who want to engage with scholars in other fields, new opportunities for interdisciplinary experimentation and collaboration" (1999, p. 453).

Thus, suggests L. Richardson, the contemporary postmodernist context in which qualitative researchers now work is a propitious one: "It provides an opportunity for us to review, critique, and re-vision writing" (2000, p. 936). For her, the opportunities for writing worthy texts—books and articles that are "good reads"—are now multiple, exciting, and demanding. But, Richardson adds, "the work is harder. The guarantees are fewer. There is a lot more for us to think about" (p. 936; see also chapter 1).

Indeed, at a basic level there are risks involved, given that new forms of representation and textual experimentation will not be welcomed with open arms by all members of the research community. For example, Sanders is scathing in his comments:

> I would like to see fewer artsy-craftsy literary exercises presented as ethnographies. While occasionally interesting, these sorts of works are often fairly mediocre theatre pieces, poetry, or somewhat overwrought autobiographical sad stories about how hard things are or were for the writer. Most of us do not live interesting enough lives or write with sufficient literary skill to create the sorts of creative and autobiographical materials that prompt readers to respond with much more than squeamish embarrassment.
>
> *Sanders, 1995, p. 627*

As Eisner points out, "There is still a good deal of prejudice out there, especially for forms of qualitative research that do not look like conventional ethnography" (2001, p. 143). Perhaps, the views expressed by Sanders (1999) might be symptomatic of a backlash to come against more personalized, emotional, and evocative forms of representation, as predicted by Rinehart (1998a). Certainly, new writing practices are challenging, and they can appear threatening to more traditionally orientated researchers. When the boundaries of the scientific or realist tale are challenged, the tendency is for counterforces to mobilize. The hegemonic order fights back, attempting to police, discipline, and control new writers (see L. Richardson, 1996a, 1997; Sparkes, 1991; chapters 5, 8, and 9 in this book).

According to Denzin, it has become certifiably chic to criticize those who experiment. He notes that for some critics the verdict is already in: "The old better than the new can do the work of sociology. So forget all this experimental stuff" (1996, p. 525). This creates a threatening environment for young scholars to work in. As Tierney and Lincoln comment, "An untenured professor, for example, obviously will find it easier to publish an article written in a standard format than one that is a performance text or a short story" (1997, p. xvi). In view of this, L. Richardson states, "One thing for us to think about is whether writing CAP [Creative Analytical Practices] ethnography for publication is a luxury open only to those who have academic sinecure" (2000, p. 936). Accordingly, undergraduate and postgraduate students, as well as untenured staff, will require support and encouragement from advisors, committees, and colleagues if they are to explore new forms of representation.

Engaging in new writing practices can be risky, but the task is not impossible. Different kinds of tales are being produced, often by young scholars, and they are being published. Therefore, those who believe that different forms of representation have something positive and worthwhile to offer should take courage from the advances that have been made in this area and not be deterred from experimenting. Indeed, they should see themselves as making a valuable contribution to the ongoing development and strengthening of qualitative research in sport and physical activity. Universities, as Eisner argues, need to encourage this development:

> I hope that in universities there will be opportunities for students to develop the skills needed to use new forms of representation to conduct qualitative research. To work effectively in the arts at least four human abilities are critical. You need refined sensibilities, you need an idea, you need imagination, and you need technical skills. Without refined sensibilities the ability to read the subtleties of the world, including the subtleties of one's own work, is impaired. Without an idea of importance whatever is created is likely to be trivial. Without imagination the work produced will be pedestrian, unable to catch and hold the reader's interest. Without technical skills all the sensibilities, ideas and imagination in the world will remain hopeless aspiration. While these qualities are critical in the arts, they are also important for doing good qualitative research.

> If we cannot see the situations we look at, we will have nothing to say about them. If we don't have an idea that matters, what we say will not be worth reading. If we can't use our imagination to give it form, it will not capture the reader's attention, and if we don't have

the technical skills to work with the constraints and affordances of a medium, our intentions will go unrealized. The good news is that these abilities can be developed. Universities need the appetite to do so.

Eisner, 2001, pp. 143-144

If the abilities described by Eisner (2001) are important for conducting qualitative research, then experimenting with different forms of representation has the potential to act as a stimulus for honing these abilities. Such a pedagogical opportunity should not be missed.

Furthermore, given the novelty of the various writing practices that are now emerging, any attempt to close down the conversation about telling different tales seems somewhat premature. Most certainly, it is far too early to pass verdict in the field of sport and physical activity, where new writing practices require time, patience, and nurturing to prove their worth. Of course, judgments will be made along the way—hopefully using appropriate criteria!

Not surprisingly, Tierney argues, "undoubtedly some experiments in narrative will fail, but others will succeed and, in doing so, they will enable us to see the world in dramatically different ways" (1993, p. 314). In view of this, perhaps the risks are worth taking even though we don't know where this experimentation will eventually take us.

With regard to the development of writing as a creative analytical practice and as a way of knowing, L. Richardson (2000) suggests exercises that encourage students to experiment with new forms of writing in ways that attempt to demystify the process, nurture the researcher's voice, and serve the process of discovery. These include transforming field notes into drama, transforming an in-depth interview into a poetic representation, experimenting with autoethnography, writing a layered text, writing the same data in three different ways, writing stories, and experimenting with different forms of writing for different audiences and different occasions.

With each of the suggested exercises, L. Richardson (2000) includes important questions that focus attention on how the author is coming to know differently through the use of a different genre. For example, with regard to transforming an in-depth interview into a poetic representation, she asks, "Where are you in the poem? What do you know about the interviewee and about yourself that you did not know before you wrote the poem? What poetic devices have you sacrificed in the name of science?" (p. 942). There is a clear sense of purpose to Richardson's calls for experimentation; she is not advocating "art for art's sake."

Here, the concerns expressed by Eisner (2001) about some of the efforts he has encountered to invent new approaches to education research are worthy of our attention. In particular, he is concerned about the connection or lack thereof between the form a research project takes and the degree to which it informs someone about something. For Eisner, one of the virtues of propositional discourse is that it has the capacity, when well used, to describe situations in reasonably precise terms. Of course, as he quickly points out, it is not free from ambiguity, but it can be used in ways that promote mutual understanding.

> By contrast, I have seen at conferences presentations conceived of as examples of the new wave in qualitative research that appeared to me to have more to do with novelty than with an effort to inform. One presentation I saw used a coffin accompanied by a bevy of pallbearers to illustrate a theme whose point escaped me. The presentation was novel, the image vivid, but in the end uninformative. It is critical that there be sufficient clarity to render a work useful to someone. Put another way, researchers who employ inventive ways of presenting what has been learned have an obligation to create something that a reader or viewer will find meaningful.
>
> *Eisner, 2001, p. 139*

Accordingly, researchers of any paradigmatic persuasion in sport and physical activity should pause for thought before rushing out to produce experimental texts in the spirit of noncritical textual radicalism. As Coffey and Atkinson point out, while we need to be aware of the variety of strategies now available, "We should not, however, experiment simply for the sake of experimentation" (1996, p. 137). Essentially, as P. Atkinson argues, just as we must take responsibility for theoretical and methodological decisions, so textual or representational decisions must be made responsibly:

> There is no need for sociologists all to flock towards "alternative" literary modes. The point of the argument is not to suggest that suddenly, from now on, sociological ethnography should be represented through pastiche or literary forms. The discipline will not be aided by the unprincipled adoption of any particular textual practices, "literary" or otherwise. On the other hand, we must always be aware that there are many available styles.
>
> *P. Atkinson, 1990, p. 180*

The unprincipled and noncritical adoption of new forms of representation would, indeed, be disastrous for research in sport and physical activity. For example, it might well lead to one orthodoxy being replaced by another. Here, qualitative researchers would be pressured to report their findings as ethnodramas or poetic representations rather than realist tales, regardless of their intentions and purposes. Unprincipled adoption would also deflect attention away from other serious issues.

For example, we need to consider not only the problems associated with the creation of the final product but also the relationship between this final product and the research process itself. On this issue, Agar (1995) expresses concern over the general lack of discussion that has taken place regarding the research process and how it would need to change to be in harmony with the final products. He feels that there is a danger that qualitative researchers, in their newly awakened enthusiasm to focus on the text as a neglected product, might engage in what Clifford (1986) calls a *fetishizing of form* and thereby lose sight of the *process* side of what they do. When this happens there is a tendency to define process and product as separate problems when in fact they are intimately related to each other. Agar goes on to point out that one unintended consequence of this dislocation is that theories of text can develop in isolation from the research processes whose results they supposedly represent. Having highlighted these and other issues he concludes:

> Textuality, as a consciousness-raising concept, is long overdue. But textuality, as the primary focus for what ethnography is all about, is, I think, a mistake. When process and product tug against each other, ethnographic credibility turns sour—credibility in the sense of making a good argument that displays and accounts for samples of group life. At a time when interest in ethnographic research is growing, when our sense of what it is and how it works is improving, a move to new textual forms without more attention to the research processes that ground them would be a serious ethno-mistake.
>
> *Agar, 1995, pp. 128-129*

Others have also warned that a preoccupation with style and genre runs the risk of aestheticism in which the value of writing about research exhausts itself in the pleasure of the text (Geertz, 1988; Rosaldo, 1993). These timely words of warning guard us against substituting one form of neglect (for the product) for another (the process). After all, as Van Maanen (1995a) points out, textual analysis alone does not provide a better way to do ethnography in the field even though it does remind us of the limits of representational possibilities. Consequently, researchers need to be constantly aware of the intimate relationship between the process and product when they experiment with new forms of writing.

Another problem associated with aestheticism is the risk of elevating style, or panache, over content. As Wolcott (1995) is quick to point out, no amount of style can cover for a problem with focus. Certainly, style is necessary, but it is not sufficient in and of itself. For Wolcott, a concern for wordsmithing should not take over, "so that style becomes a preoccupation, leading to 'slick description' instead of 'thick.' Efforts at good writing need to be coupled with having something worthwhile to say" (p. 218). I think even the most energetic advocate of new forms of representation in sport and physical activity, myself included, would agree with this view.

Given the situation described above, we need to proceed with enthusiasm but also with care and caution. As Eisner suggests, "We need to walk the line between the risk inherent in innovation and the need to do work that has the quality it needs to persuade" (2001, p. 143). It seems likely that, for the moment at least, scientific tales will continue as the dominant form of representation for positivist and postpositivist researchers, and realist tales will be the main ones produced by qualitative researchers. Putting aside the risks of not getting published if one deviates from this kind of telling, this dominance is not surprising. These tales have served us well in the past for particular purposes, and they will continue to contribute to our understanding in the future. They are certainly not to be dismissed lightly despite a growing awareness of their limitations in specific contexts and for certain purposes.

Experimental forms of writing are still not the standard style and therefore will not be for everyone. Furthermore, as P. Atkinson et al. note, "Recent innovations do not have to be seen as wholesale rejections of prior positions" (1999, p. 470). Likewise, Anderson emphasizes that the new ethnographic genres do not necessarily signal the demise of previously existing ones, "but rather add more options in the styles of research and analysis available to qualitative researchers" (1999, p. 453). In this sense, it is more sensible to speak of the *displacement* and *relocation* of realist and scientific tales than their replacement. With regard to the discipline of anthropology, Rosaldo expresses the following view:

> Rather than discarding distanced normalizing accounts, the discipline should recover them, but with a difference. They must be cut down to size and relocated, not replaced. No longer enshrined as ethnographic realism, the sole vehicle for speaking the literal truth about other cultures, the classic norms should become one mode of representation among others. . . . an increased disciplinary tolerance for diverse legitimate rhetorical forms will allow for any particular text to be read against other possible versions. Allowing forms of writing that have been marginalized or banned altogether to gain legitimacy could enable the discipline to approximate people's lives from angles of vision. Such a tactic would better enable us to advance the ethnographic project of apprehending the range of human possibilities in their fullest complexities.
>
> *Rosaldo, 1993, p. 62*

Textual experimentation will not eradicate more conventional forms of reporting, nor should it. What this experimentation offers is a wider range of choice, not a narrowing of selections. Accordingly, Woods situates his own writing between traditional realist tales and experimental writing and adds, "I see no reason to abandon established criteria of worth in academic presentation, but I do feel that some of these new forms promise to reach parts of social understanding that established methods cannot reach" (1999, p. 48). This position is supported by L. Richardson (2000), who argues that writing in traditional ways does not prevent researchers from writing in other ways for other audiences at different times.

Furthermore, L. Richardson (2000) emphasizes, researchers will not lose the language of science should they choose to learn to write in other ways, any more than students who learn a second language lose their native tongue. Rather, as she points out, acquiring a second language can enrich the lives of students in two ways: "It gains them entry into a new culture and literature, and it leads them to a deepened understanding of their first language, not just grammatically, but as a language that constructs how they view the world" (p. 936).

Increasing the repertoire of tales told gives qualitative researchers both the luxury and the problem of choice. As Tierney (1993) argues, not all researchers need choose to construct ethnographic fictions since their *intentions* and *purposes* may necessitate a more traditional case study and form of reporting. Tierney suggests, however, that there might be times when we "need to create texts that enable the reader to reflect on his or her own life and see if the text resembles any sense of reality" (p. 313). In such circumstances, the author might consider producing an ethnographic fiction as opposed to a scientific or realist tale. Likewise, those who wish to communicate, for example, the deep emotional experiences associated with involvement in sport and physical activity might choose an ethnographic drama, an autoethnography, or perhaps a poetic representation. They might also choose to move into creative fiction if they feel this genre is best suited to this task.

Indeed, it may well be that some experiences can only find their expression in specific kinds of tales. As Verma argues, although there has been a blurring of genres in literature, certain distinctions still exist between a story, a poem, and a novel:

> Thus the choice of forms cannot be arbitrary; it *is inherent in the nature of experience itself.* We cannot transfer the same experience from one form to another without deadening its quick throb. Once dead, it can be transferred anywhere, to a play or a poem or a story. It is not that the writer first has a certain experience and then he embodies it in particular art form, *rather it is the experience which chooses its own form to make its presence felt.* Thus we cannot say that a particular experience has been "captured" in a story; there is no capturing in

the realm of art. What is more correct is that a *certain experience could only realise itself in the form of a story, and in no other form.*

Verma, 1991, p. 6 (my italics)

If, in many ways, the experience does in part choose the form of representation to make its presence felt, then—given the multidimensionality of experiences in sport and physical activity—qualitative researchers in sport and physical activity may have no option but to tell different tales. Furthermore, in these conditions, making an informed choice about the form becomes even more crucial to the enterprise. Of course, active authorial decisions based on informed choice require an awareness of the many ways in which the findings of qualitative research can be turned into texts and other forms of output. Unfortunately, this is rarely the case and, too often in the past, the decision about what kind of tale to tell has been a passive one, a default choice made because the author was not aware of the options available.

This lack of awareness is likely to change in the coming years as we become increasingly conscious of how our "tales from the field" are constructed. Already, the ways we write about ourselves and others have become problematic, and we cannot ignore this central aspect of our work. Our texts can no longer be taken to be innocent or neutral. In view of this, P. Atkinson (1991) suggests that we need to develop a *reflexive self-awareness* of the rhetorical and stylistic conventions we use, not to substitute textual analysis for fieldwork but rather to bring these concepts within our explicit and methodological understanding. He emphasizes that such reflexiveness is not easy since it requires an acquaintance with recent and contemporary literary theory along with parallel work on the poetics of economics, history, law, and so on. He also recognizes that it is far easier to copy a taken-for-granted model than it is to understand other genres, to manipulate their conventions, and to experiment with them. In summarizing the situation, Coffey and Atkinson comment:

> The growing awareness among social scientists of the significance of representations, and of their various forms, means *that there can be no excuse for failure to be conscious of them.* We do not want to preach one or another approach to writing and representation, traditional or experimental. We do, however, wish to commend an awareness of variety, coupled with principled decisions.

Coffey & Atkinson, 1996, p. 137 (my italics)

My hope is that in the coming years a variety of representational forms will come to be valued in their own right for the powerful ways in which they can enhance and extend our understanding of sport and physical activity. Their acceptance will, of course, not be unconditional, and like all tellings they will have to prove their worth. This is as it should be, as long as the research community is prepared to give new writing practices a fair and just hearing. I am confident that it will, as there is so much to be gained from expanding our horizons and engaging with other ways of knowing. Where all this will take us, I don't know. Like L. Richardson, however, "I do know that the ground has been staked, the foundation laid, the scaffolding erected, and diverse and adventurous settlers have moved in" (2000, p. 939). Chances have been taken by scholars in sport and physical activity, and they have created intellectual spaces for others to move into and explore. Moving into unknown space can be frightening, but it can also be exhilarating. Sometimes it can be both. Only you can decide.

References

Adams, N., Causey, T., Jacobs, M.-E., Munro, P., Quinn, M., & Trousdale, A. (1998). "Womentalkin." *Qualitative Studies in Education, 11*(3), 383-395.

Agar, M. (1995). Literary journalism as ethnography. In J. Van Maanen (Ed.) *Representation in ethnography* (pp. 112-129). London: Sage.

American Psychological Association. (1994). *Publication manual of the American Psychological Association* (4th ed.). Washington, DC: Author.

Anderson, L. (1999). The open road to ethnography's future. *Journal of Contemporary Ethnography, 28*(5), 451-459.

Angrosino, M. (1998). *Opportunity House*. London: Altamira Press.

Athens, L. (1995). Dramatic self change. *The Sociological Quarterly, 36*(30), 571-586.

Atkinson, P. (1990). *The ethnographic imagination*. London: Routledge.

Atkinson, P. (1991). Supervising the text. *International Journal of Qualitative Studies in Education, 4*(2), 161-174.

Atkinson, P. (1992). *Understanding ethnographic texts*. London: Sage.

Atkinson, P., Coffey, A., & Delamont, S. (1999). Ethnography. *Journal of Contemporary Ethnography, 28*(5), 460-471.

Atkinson, P., Coffey, A., & Delamont, S. (2001). Editorial. *Qualitative Research, 1*(1), 5-21.

Atkinson, R. (1998). *The life story interview*. London: Sage.

Austin, D. (1996). Kaleidoscope. In C. Ellis & A. Bochner (Eds.), *Composing ethnography* (pp. 206-230). London: Altamira Press.

Baff, S. (1997). "Realism and naturalism and dead dudes." *Qualitative Inquiry, 3*(4), 469-490.

Bain, L. (1992). Research in sport pedagogy. In T. Williams, L. Almond, & A. Sparkes (Eds.), *Sport and physical activity* (pp. 3-22). London: E & FN Spon.

Bandy, S., & Darden, A. (Eds.). (1999a). *Crossing boundaries*. Champaign, IL: Human Kinetics.

Bandy, S., & Darden, A. (Eds.). (1999b). Prelude. In S. Bandy & A. Darden (Eds.), *Crossing boundaries* (pp. ix-xiv). Champaign, IL: Human Kinetics.

Banks, A., & Banks, S. (1998). The struggle over facts and fictions. In A. Banks & S. Banks (Eds.), *Fiction and social research* (pp. 11-29). London: Altamira Press.

Banks, S. (2000). Five holiday letters. *Qualitative Inquiry, 6*(3), 392-405.

Barone, T. (1990). Using the narrative text as an occasion for conspiracy. In E. Eisner & A. Peshkin (Eds.), *Qualitative inquiry in education* (pp. 305-326). New York: Teachers College Press.

Barone, T. (1995). Persuasive writings, vigilant readings, and reconstructed characters. *Qualitative Studies in Education, 8*(1), 63-74.

Barone, T. (1997). Among the chosen. *Qualitative Inquiry, 3*(2), 222-236.

Barone, T. (2000). *Aesthetics, politics, and educational inquiry*. New York: Peter Lang.

Becker, H. (1986). *Writing for social scientists*. Chicago: University of Chicago Press.

Bernstein, R. (1983). *Beyond objectivism and relativism*. Oxford: Basil Blackwell.

Bernstein, R. (1991). *The new constellation*. Cambridge, UK: Polity Press.

Best, J. (1995). Lost in the ozone again. *Studies in Symbolic Interaction, 17*, 125-130.

Bethanis, P. (2000). The shadowboxer. *Sociology of Sport Journal, 17*(1), 81-82.

Blumenfeld-Jones, D. (1995). Fidelity as a criterion for practising and evaluating narrative inquiry. In J. Hatch & R. Wisniewski (Eds.), *Life history and narrative* (pp. 25-35). London: Falmer Press.

References

Bochner, A. (1994). Perspectives on inquiry II. In M. Knapp & G. Miller (Eds.), *Handbook of interpersonal communication* (pp. 21-41). London: Sage.

Bochner, A. (1997). It's about time. *Qualitative Inquiry, 3*(4), 418-438.

Bochner, A. (2001). Narrative's virtues. *Qualitative Inquiry, 7*(2), 131-157.

Bochner, A., & Ellis, C. (1996a). Taking ethnography into the twenty-first century. *Journal of Contemporary Ethnography, 25*(1), 3-5.

Bochner, A., & Ellis, C. (1996b). Talking over ethnography. In C. Ellis & A. Bochner (Eds.), *Composing ethnography* (pp. 13-45). London: Altamira Press.

Bochner, A., & Ellis, C. (2002). *Ethnographically speaking.* New York: Altamira Press.

Bochner, A., & Waugh, J. (1995). Talking-with as a model for writing-about. In L. Langsdorf & A. Smith (Eds.), *Recovering pragmatism's voice* (pp. 211-223). Albany, NY: SUNY Press.

Boman, J., & Jevne, R. (2000). Ethical evaluation in qualitative research. *Qualitative Health Research, 10*(4), 547-554.

Brackenridge, C. (1999). Managing myself. *International Review for the Sociology of Sport, 34*(4), 399-410.

Brackenridge, C. (2001). *Spoilsports.* London: Routledge.

Brady, I. (1998). A gift of the journey. *Qualitative Inquiry, 4*(4), 463.

Brady, I. (1999). Two poems. *Qualitative Inquiry, 5*(4), 566-567.

Brady, I. (2000). Three Jaguar/Mayan intertexts. *Qualitative Inquiry, 6*(1), 58-64.

Brown, L. (1998). "Boys' training." In C. Hickey, L. Fitzclarence, & R. Matthews (Eds.), *Where the boys are* (pp. 83-96). Geelong, Australia: Deakin University Press.

Bruce, T. (2002). Pass. In J. Denison & P. Markula (Eds.), *"Moving writing"* (pp. 129-144). New York: Peter Lang.

Bruner, J. (1986). *Actual minds, possible worlds.* Cambridge, MA: Harvard University Press.

Carroll, L. (1872). *Through the looking glass and what Alice found there.* London: Macmillan.

Ceglowski, D. (1997). That's a good story, but is it really research? *Qualitative Inquiry, 3*(20), 188-201.

Ceglowski, D. (2000). Research as relationship. *Qualitative Inquiry, 6*(2), 88-103.

Charmaz, K. (1987). Struggling for a self. In J. Roth & P. Conrad (Eds.), *Research in the sociology of health care: Vol. 6. The experience and management of chronic illness.* Greenwich, CT: JAI Press.

Charmaz, K., & Mitchell, R. (1997). The myth of silent authorship. In R. Hertz (Ed.), *Reflexivity and voice* (pp. 193-215). London: Sage.

Cherryholmes, C. (1988). *Power and criticism.* New York: Teachers College Press.

Cheyne, A., & Tarulli, D. (1998). Paradigmatic psychology in narrative perspective. *Narrative Inquiry, 8*(1), 1-25.

Christensen, P. (2000). Believing. *Sociology of Sport Journal, 17*(1), 83-94.

Church, K. (1995). *Forbidden narratives.* London: Gordon & Breach.

Clarke, G., & Humberstone, B. (Eds.). (1997) *Researching women and sport.* London: Macmillan Press.

Clifford, J. (1983). On ethnographic authority. *Representations, 1,* 118-146.

Clifford, J. (1986). Introduction. In J. Clifford & G. Marcus (Eds.), *Writing culture* (pp. 1-26). Los Angeles: University of California Press.

Clough, P. (1999). Crises of schooling and the "crisis of representation." *Qualitative Inquiry, 5*(3), 428-488.

Clough, P. (2000). A familial unconsciousness. *Qualitative Inquiry, 6*(3), 318-336.

Coe, D. (1991). Levels of knowing in ethnographic inquiry. *International Journal of Qualitative Studies in Education, 4,* 313-331.

Coffey, A. (1999). *The ethnographic self.* London: Sage.

Coffey, A., & Atkinson, P. (1996). *Making sense of qualitative data.* London: Sage.

Cole, C. (1991). The politics of cultural representation. *International Review for the Sociology of Sport, 26*(1), 36-49.

Coles, R. (1989). *The call of stories.* Boston: Houghton Mifflin.

References

Delamont, S., Coffey, A., & Atkinson, P. (2000). The twilight years? *International Journal of Qualitative Studies in Education, 13*(3), 223-238.

Denison, J. (1994). *Sport retirement*. Unpublished doctoral dissertation, University of Illinois, Urbana.

Denison, J. (1996). Sport narratives. *Qualitative Inquiry, 2*(3), 351-362.

Denison, J. (1999). Boxed in. In A. Sparkes & M. Silvennoinen (Eds.), *Talking bodies* (pp. 29-36). Jyvaskyla, Finland: SoPhi.

Denison, J. (2000). Gift. *Sociology of Sport Journal, 17*(1), 98-99.

Denison, J., & Markula, P. (2002). Introduction. In J. Denison & P. Markula (Eds.), *"Moving writing"* (pp. 1-22). New York: Peter Lang.

Denison, J., & Rinehart, R. (2000). Introduction. *Sociology of Sport Journal, 17*(1), 1-4.

Denzin, N. (1994). The art and politics of interpretation. In: N. Denzin & Y. Lincoln (Eds.), *Handbook of qualitative research* (pp. 500-515). London: Sage.

Denzin, N. (1996). Punishing poets. *Qualitative Sociology, 19*(4), 525-528.

Denzin, N. (1997). *Interpretive ethnography*. London: Sage.

Denzin, N. (2000). The practices and politics of interpretation. In N. Denzin & Y. Lincoln (Eds.), *Handbook of qualitative research* (2nd ed., pp. 897-922). London: Sage.

Denzin, N., & Lincoln, Y. (1994). Introduction. In N. Denzin & Y. Lincoln (Eds.), *Handbook of qualitative research* (pp. 1-17). London: Sage.

Denzin, N., & Lincoln, Y. (2000). Introduction. In N. Denzin & Y. Lincoln (Eds.), *Handbook of qualitative research* (2nd ed., pp. 1-29). London: Sage.

DeVault, M. (1996). In defense of textual experimentation. *Qualitative Sociology, 19*(4), 529-531.

DeVault, M. (1997). Personal writing in social research. In R. Hertz (Ed.), *Reflexivity and voice* (pp. 216-228). London: Sage.

Dévis, J., & Sparkes, A. (1999). Burning the book. *European Physical Education Review, 5*(2), 135-152.

Dewar, A. (1990). Oppression and privilege in physical education. In D. Kirk & R. Tinning (Eds.), *Physical education, curriculum and culture* (pp. 67-99). London: Falmer Press.

Dewar, A. (1991, August). Feminist pedagogy in physical education. *Journal of Physical Education, Recreation and Dance,* 68-77.

Dewar, A. (1993). Will all the generic women in sport please stand up? *Quest, 45,* 211-229.

Diversi, M. (1998). Glimpses of street life. *Qualitative Inquiry, 4*(2), 131-147.

Donmoyer, R., & Donmoyer, J. (1995). Data as drama. *Qualitative Inquiry, 1*(4), 402-428.

Donmoyer, R., & Donmoyer, J. (1998). Reader's theater and educational research—give me a for instance. *Qualitative Studies in Education, 11*(3), 397-407.

Dowling Naess, F. (1998). *Tales of Norwegian physical education teachers*. Unpublished doctoral dissertation, The Norwegian University of Sport and Physical Education, Oslo, Norway.

Dunbar, C. (1999). Three short stories. *Qualitative Inquiry, 5*(1), 130-140.

Dunbar, C. (2001). From alternative school to incarceration. *Qualitative Inquiry, 7*(2), 158-170.

Duncan, M.C. (1998). Stories we tell ourselves about ourselves. *Sociology of Sport Journal, 15,* 95-108.

Duncan, M.C. (2000). Reflex. *Sociology of Sport Journal, 17*(10), 60-68.

Duquin, M. (2000). Sport and emotions. In J. Coakley & E. Dunning (Eds.), *Handbook of sport sociology* (pp. 477-489). London: Sage.

Eakin, P. (1999). *How our lives become stories*. Ithaca, NY: Cornell University Press.

Eisner, E. (1988). The primacy of experience and the politics of method. *Educational Researcher, 29,* 99-113.

Eisner, E. (1991). *The enlightened eye*. New York: Macmillan.

Eisner, E. (1997). The new frontier in qualitative research methodology. *Qualitative Inquiry, 3*(3), 259-273.

References

Eisner, E. (2001). Concerns and aspirations for qualitative research in the new millennium. *Qualitative Research, 1*(2), 135-145.

Ellis, C. (1993). "There are survivors." *The Sociological Quarterly, 34*, 711-730.

Ellis, C. (1995a). *Final negotiations*. Philadelphia: Temple University Press.

Ellis, C. (1995b). The other side of the track. *Qualitative Inquiry, 1*(2), 147-167.

Ellis, C. (1995c). Speaking of dying. *Symbolic Interaction, 18*, 73-81.

Ellis, C. (1997). Evocative autoethnography. In W. Tierney & Y. Lincoln (Eds.), *Representation and the text* (pp. 115-139). New York: State University of New York Press.

Ellis, C. (1998) "I hate my voice." *The Sociological Quarterly, 39*(4), 517-537.

Ellis, C. (1999). Heartful autoethnography. *Qualitative Health Research, 9*(5), 669-683.

Ellis, C. (2000) Creating criteria. *Qualitative Inquiry, 6*(2), 273-277.

Ellis, C. (2001). With mother/with child. *Qualitative Inquiry, 7*(5), 598-616.

Ellis, C., & Bochner, A. (1992). Telling and performing personal stories. In: C. Ellis & M. Flaherty (Eds.), *Investigating subjectivity* (pp. 79-101). London: Sage.

Ellis, C., & Bochner, A. (Eds.). (1996). *Composing ethnography*. London: Altamira Press.

Ellis, C., & Bochner, A. (2000). Autoethnography, personal narrative, reflexivity. In N. Denzin & Y. Lincoln (Eds.), *Handbook of qualitative research* (2nd ed., pp. 733-768). London: Sage.

Ellis, C., & Flaherty, M. (Eds.). (1992). *Investigating subjectivity*. London: Sage.

Ellsworth, E. (1992). Why doesn't this feel empowering? In C. Luke & J. Gore (Eds.), *Feminisms and critical pedagogy* (pp. 90-119). London: Routledge.

Ely, M., Vinz, R., Downing, M., & Anzul, M. (1997). *On writing qualitative research*. London: Falmer Press.

Evans, J. (1992). A short paper about people, power and educational reform. In A. Sparkes (Ed.), *Research in physical education and sport* (pp. 231-247). London: Falmer Press.

Faulkner, G., & Sparkes, A. (1999). Exercise as therapy for schizophrenia. *Journal of Sport & Exercise Psychology, 21*(1), 52-69.

Featherstone, M. (1991). *Consumer culture and postmodernism*. London: Sage.

Fernandez-Balboa, J.-M. (1998). Transcending masculinities. In C. Hickey, L. Fitzclarence, & R. Matthews (Eds.), *Where the boys are* (pp. 121-139). Geelong, Australia: Deakin University Press.

Feyerabend, P. (1987). *Farewell to reason*. London: Verso.

Fine, A. (1999). Field labor and ethnographic reality. *Journal of Contemporary Ethnography, 28*(5), 532-539.

Fine, M. (1994). Working the hyphens. In N. Denzin & Y. Lincoln (Eds.), *Handbook of qualitative research* (pp. 70-82). London: Sage.

Fine, M., & Weis, L. (1998). Writing the "wrongs" of fieldwork. In J. Smyth & G. Shacklock (Eds.), *Being reflexive in critical educational and social research* (pp. 13-35). London: Falmer.

Fine, M., Weis, L., Weseen, S., & Wong, L. (2000). In N. Denzin & Y. Lincoln (Eds.), *Handbook of qualitative research* (2nd ed., pp. 107-132). London: Sage.

Finley, S. (2000). "Dream child." *Qualitative Inquiry, 6*(3), 432-434.

Finley, S., & Finley, M. (1999). Sp'ange. *Qualitative Inquiry, 5*(3), 313-337.

Foley, D. (1992). Making the familiar strange. *Sociology of Sport Journal, 9*(1), 36-47.

Foucault, M. (1977). *Discipline and punish, the birth of the person*. London: Tanstock.

Frank, A. (1991). *At the will of the body*. Boston: Houghton Mifflin.

Frank, A. (1995). *The wounded storyteller*. Chicago: University of Chicago Press.

Frank, K. (2000). "The management of hunger." *Qualitative Inquiry, 6*(4), 474-488.

Freeman, M. (1993). *Rewriting the self*. London: Routledge.

Fussell, S. (1991). *Muscle*. London: Scribners.

Gamson, J. (2000). Sexualities, queer theory, and qualitative research. In N. Denzin, & Y. Lincoln (Eds.), *Handbook of qualitative research* (pp. 347-365). London: Sage.

Garratt, D., & Hodkinson, P. (1998). Can there be criteria for selecting research criteria? *Qualitative Inquiry, 4*(4), 515-539.

Geertz, C. (1974). *The interpretation of cultures*. New York: Basic Books.

Geertz, C. (1984). Anti anti-relativism. *American Anthropologist, 86*, 263-278.

Geertz, C. (1988). *Works and lives.* Cambridge: Polity Press.

Gergen, K. (1999). *An invitation to social construction.* London: Sage.

Gergen, M., & Gergen, K. (2000). Qualitative inquiry. In N. Denzin & Y. Lincoln (Eds.), *Handbook of qualitative research* (2nd ed., pp. 1025-1046). London: Sage.

Glesne, C. (1997). That rare feeling. *Qualitative Inquiry, 3*(2), 202-221.

Golden-Biddle, K., & Locke, K. (1997). *Composing qualitative research.* London: Sage.

Goldstein, T. (2001). Hong Kong, Canada. *Qualitative Inquiry, 7*(3), 279-303.

Gore, J. (1990). Pedagogy as text in physical education teacher education. In D. Kirk & R. Tinning (Eds.), *Physical education, curriculum and culture* (pp.101-138). Lewes, UK: Falmer Press.

Gore, J. (1992). What can we do for you! What *can* "we" do for "you"? In C. Luke & J. Gore (Eds.), *Feminisms and critical pedagogy* (pp. 54-73). London: Routledge.

Gorelick, S. (1991). Contradictions of feminist methodology. *Gender & Society, 5*(4), 459-477.

Gottschalk, S. (1998). Postmodern sensibilities and ethnographic possibilities. In A. Banks & S. Banks (Eds.), *Fiction and social research* (pp. 205-233). London: Altamira Press.

Gould, D., Udry, E., Bridges, D., & Beck, L. (1997). Coping with season-ending injuries. *The Sport Psychologist, 11*, 379-399.

Gray, R. (2000a). Graduate school never prepared me for this. *Reflective Practice, 1*(3), 377-390.

Gray, R. (2000b). *Legacy.* Harriman, TN: Men's Studies Press.

Gray, R., Greenberg, M., Fitch, M., Hamson, A., & Labrecque, M. (1998). Information needs of women with metastatic breast cancer. *Cancer Prevention & Control, 2*(20), 57-62.

Gray, R., Ivonoffski, V., & Sinding, C. (2002). Making a mess and spreading it around. In A. Bochner & C. Ellis (Eds.), *Ethnographically speaking* (pp. 57-75). California: Altamira Press.

Gray, R., Sinding, C., & Fitch, M. (2001). Navigating the social context of metastatic breast cancer. *Health, 5*, 233-248.

Gray, R., Sinding, C., Ivonoffski, V., Fitch, M., Hamson, A., & Greenberg, M. (2000). The use of research-based theatre in a project related to metastatic breast cancer. *Health Expectations, 3*, 137-144.

Guba, E. (Ed.). (1990). *The paradigm dialog.* London: Sage.

Gusfield, J. (1981). The literary rhetoric of science. *American Sociological Review, 41*, 16-34.

Habermas, J. (1971). *Knowledge and human interests* (J.J. Shapiro, Trans.). Boston: Beacon Press. (Original work published in 1968.)

Halas, J. (2001). Shooting hoops at the treatment center. *Quest, 53*, 77-96.

Hall, A. (1996). *Feminism and sporting bodies.* Champaign, IL: Human Kinetics.

Hammersley, M. (1992). *What's wrong with ethnography?* London: Routledge.

Hanson, T. (1992). The mental aspects of hitting in baseball. *Contemporary Thought on Performance Enhancement, 1*(1), 49-70.

Harvey, D. (1990). *The condition of postmodernity.* Oxford: Blackwell.

Hastrup, K. (1992). Writing ethnography. In J. Okely & H. Callaway (Eds.), *Anthropology & autobiography* (pp. 116-133). London: Routledge.

Hertz, R. (1997). Introduction. In R. Hertz (Ed.), *Reflexivity and voice* (pp. vii-xviii). London: Sage.

Hollinger, R. (1994). *Postmodernism and the social sciences.* London: Sage.

Holt, N., & Sparkes, A. (2001). An ethnographic study of cohesiveness in a college soccer team over a season. *The Sport Psychologist, 15*(3), 237-259.

Hones, D. (1998). Known in part. *Qualitative Inquiry, 4*(2), 225-248.

Hooks, B. (1991). Narratives of struggle. In P. Mariani (Ed.), *Critical fictions* (pp. 53-61). Seattle: Bay Press.

Humberstone, B. (1997). Challenging dominant ideologies in the research process. In G. Clarke & B. Humberstone (Eds.), *Researching women and sport* (pp. 199-213). London: Macmillan.

Innanen, M. (1999). Secret life in the culture of thinness. In A. Sparkes & M. Silvennoinen (Eds.), *Talking bodies* (pp. 120-134). Jyvaskyla, Finland: SoPhi.

Ivonoffski, V., & Gray, R. (1999). *Handle with care?* Toronto: Toronto-Sunnybrook Regional Cancer Centre.

Jackson, D. (1990). *Unmasking masculinity.* London: Unwin Hyman.

Jackson, D. (1999). Boxing glove. In A. Sparkes & M. Silvennoinen (Eds.), *Talking bodies* (p. 48). Jyvaskyla, Finland: SoPhi.

Janesick, V. (2001). Intuition and creativity. *Qualitative Inquiry, 7*(5), 531-540.

Jones, S. (1999). Torch. *Qualitative Inquiry, 5*(2), 280-304.

Josselson, R. (1993). A narrative introduction. In R. Josselson & A. Leiblich (Eds.), *The narrative study of lives* (pp. ix-xv). London: Sage.

Kaskisaari, M. (1994). The rhythmbody. *International Review for the Sociology of Sport, 29*(1), 15-23.

Kelly, P., Hickey, C., & Tinning, R. (2000). Educational truth telling in a more reflexive modernity. *British Journal of Sociology of Education, 21*(1), 111-122.

Kennelly, B. (1992). *Breathing spaces.* Newcastle Upon Tyne, UK: Bloodaxe Books.

Kiesinger, C. (1998). Portrait of an anorexic life. In A. Banks & S. Banks (Eds.), *Fiction and social research* (pp. 115-136). London: Altamira Press.

Klein, A. (1993). *Little big men.* Albany, NY: SUNY Press.

Kluge, M.A. (2001). Confessions of a beginning qualitative researcher. *Journal of Aging and Physical Activity, 9*(3), 329-335.

Kosonen, U. (1993). A running girl. In L. Laine (Ed.), *On the fringes of sport* (pp. 16-25). St. Augustin, Germany: Academia Verlag.

Krane, V. (1994). A feminist perspective on contemporary sport psychology research. *The Sport Psychologist, 8,* 393-410.

Kress, S. (1998). Can sociology be literature? *Journal of Contemporary Ethnography, 27*(2), 270-277.

Krieger, S. (1991). *Social science and the self.* New Brunswick, NJ: Rutgers University Press.

Krizek, R. (1998). Lessons. In A. Banks & S. Banks (Eds.), *Fiction and social research* (pp. 89-113). London: Altamira Press.

Ladson-Billings, G. (2000). Racialized discourses and ethnic epistemologies. In N. Denzin, & Y. Lincoln (Eds.), *Handbook of qualitative research* (pp. 257-277). London: Sage.

Laine, L. (Ed.). (1993). *On the fringes of sport.* St. Augustin, Germany: Academia Verlag.

Langer, S. (1953). *Feeling and form.* New York: Scribner's.

Lather, P. (1986). Issues of validity in openly ideological research. *Interchange, 17,* 63-84.

Lather, P. (1991). *Getting smart.* London: Routledge.

Lather, P. (1993). Fertile obsession. *The Sociological Quarterly, 34*(4), 673-693.

Lather, P. (1995). The validity of angels. *Qualitative Inquiry, 1*(1), 41-68.

Law, J., & Williams, R. (1982). Putting the facts together. *Social Studies of Science, 12*(4), 535-558.

Lax, M. (1999). *A sporting chance.* Unpublished bachelor's dissertation, Exeter University, Exeter, UK.

Leder, D. (1990). *The absent body.* Chicago: University of Chicago Press.

Lieblich, A., Tuval-Mashiach, R., & Zilber, R. (1998). *Narrative research.* London: Sage.

Lincoln, Y. (1993). I and thou. In D. McLaughlin & W. Tierney (Eds.), *Naming silenced lives* (pp. 29-47). New York: Routledge.

Lincoln, Y. (1995). Emerging criteria for quality in qualitative and interpretive research. *Qualitative Inquiry, 1*(3), 275-289.

Lincoln, Y. (1997). Self, subject, audience, text. In W. Tierney & Y. Lincoln (Eds.), *Representation and the text* (pp. 37-56). Albany, NY: SUNY Press.

Lincoln, Y., & Denzin, N. (2000). The seventh moment. In N. Denzin & Y. Lincoln (Eds.), *Handbook of qualitative research* (2nd ed., pp. 1047-1065). London: Sage.

Lincoln, Y., & Guba, E. (2000). Paradigmatic controversies, contradictions, and emerging

References

confluences. In N. Denzin & Y. Lincoln (Eds.), *Handbook of qualitative research* (2nd ed., pp. 163-188). London: Sage.

Livia, A. (1996). Daring to presume. *Feminism & Psychology, 6*(1), 31-41.

Luke, C., & Gore, J. (Eds.). (1992). *Feminisms and critical pedagogy.* London: Routledge.

Lyon, D. (1994). *Postmodernity.* Mitlon Keynes, UK: Open University Press.

Lyons, K. (1992). Telling stories from the field? In A. Sparkes (Ed.), *Research in physical education and sport* (pp. 248-270). London: Falmer Press.

Macdonald, D., & Tinning, R. (1995). Physical education teacher education and the trend to proletarianization. *Journal of Teaching in Physical Education, 15*(1), 98-118.

Manning, K. (1997). Authenticity in constructivist inquiry. *Qualitative Inquiry, 3*(1), 93-115.

Marcus, G., & Cushman, D. (1982). Ethnographies as texts. *Annual Review of Anthropology, 11,* 25-69.

Marcus, G., & Fischer, M. (Eds.). (1986). *Anthropology as cultural critique.* Chicago: University of Chicago Press.

Marcus, J. (1992). Racism, terror and the production of Australian auto/biographies. In J. Okely & H. Callaway (Eds.), *Anthropology & autobiography* (pp. 100-115). London: Routledge.

Markula, P. (2002). Bodily dialogues. In J. Denison & P. Markula (Eds.), *"Moving writing"* (pp. 25-50). New York: Peter Lang.

Markula, P., Grant, B., & Denison, J. (2001). Qualitative research and aging and physical activity. *Journal of Aging and Physical Activity, 9*(3), 245-264.

McCall, M. (2000). Performance ethnography. In N. Denzin & Y. Lincoln (Eds.), *Handbook of qualitative research* (2nd ed., pp. 420-433). London: Sage.

McLaughlin, D., & Tierney, W. (1993). *Naming silenced lives.* London: Routledge.

McLeod, J. (1997). *Narrative and psychotherapy.* London: Sage.

McTaggart, R. (1997). Reading the collection. In R. McTaggart (Ed.), *Participatory action research* (pp. 1-24). Albany, NY: SUNY Press.

Mienczakowski, J. (1995). The theater of ethnography. *Qualitative Inquiry, 1*(3), 360-375.

Mienczakowski, J. (1996). An ethnographic act. In C. Ellis & A. Bochner (Eds.), *Composing ethnography* (pp. 244-264). London: Altamira Press.

Mienczakowski, J. (2001). Ethnodrama. In P. Atkinson, A. Coffey, S. Delamont, J. Lofland, & L. Lofland (Eds.), *Handbook of ethnography* (pp. 468-476). London: Sage.

Milchrist, P. (2001). Alzheimer's. *Journal of Aging and Physical Activity, 9*(3), 265-268.

Miles, M., & Huberman, M. (1994). *Qualitative data analysis.* London: Sage.

Miller, E. (2000). Dis. *Sociology of Sport Journal, 17*(1), 75-80.

Miller, M. (1998). (Re)presenting voices in dramatically scripted research. In A. Banks & S. Banks (Eds.), *Fiction and social science* (pp. 67-78). London: Altamira Press.

Minh-ha, T. (1989). *Women, native, other.* Bloomington, IN: Indiana University Press.

Morgan, D. (1998). Sociological imaginings and imagining sociology. *Sociology, 32*(4), 647-663.

Morrison, B. (1998). *Too true.* London: Granta Books.

Mykhalovskiy, E. (1996). Reconsidering table talk. *Qualitative Sociology, 19*(1), 131-151.

Nelson, J., Megill, A., & McCloskey, D. (1987). Rhetoric of inquiry. In J. Nelson, A. Megill, & D. McCloskey (Eds.), *The rhetoric of the human sciences* (pp. 3-18). Madison, WI: University of Wisconsin Press.

Neumann, M. (1996). Collecting ourselves at the end of the century. In C. Ellis & A. Bochner (Eds.), *Composing ethnography* (pp. 172-198). London: Altamira Press.

Newman, M. (1992). Perspectives on the psychological dimension of goalkeeping. *Contemporary Thought on Performance Enhancement, 1*(1), 71-105.

Nilges, L. (1998). I thought only fairy tales had supernatural power. *Journal of Teaching in Physical Education, 17*(2), 172-194.

Nilges, L. (2001). The twice-told tale of Alice's physical life in Wonderland. *Quest, 53,* 231-259.

References

Olesen, V. (2000). Feminisms and qualitative research at and into the millennium. In N. Denzin & Y. Lincoln (Eds.), *Handbook of qualitative research* (2nd ed., pp. 215-256). London: Sage.

Oliver, K. (1998). A journey into narrative analysis. *Journal of Teaching in Physical Education, 17*(2), 244-259.

Orner, M. (1992). Interrupting the calls for student voice in "liberatory" education. In C. Luke & J. Gore (Eds.), *Feminisms and critical pedagogy* (pp. 74-89). London: Routledge

Paget, M. (1990). Performing the text. *Journal of Contemporary Ethnography, 19*(1), 136-155.

Parrott, K. (2002). Initiation. In J. Denison & P. Markula (Eds.), *"Moving writing"* (pp. 87-109). New York: Peter Lang.

Pedersen, K. (1998) Doing feminist ethnography in the "wilderness" around my hometown. *International Review for the Sociology of Sport, 33*(4), 393-402.

Pelias, R. (1999). *Writing performance.* Carbondale, IL: Southern Illinois University Press.

Pifer, D. (1999). Small town race. *Qualitative Inquiry, 5*(4), 541-562.

Plath, D. (1990). Fieldnotes, filed notes, and the conferring of note. In R. Sanjek (Ed.), *Fieldnotes* (pp. 371-384). Ithaca, NY: Cornell University Press.

Plummer, K. (2001). *Documents of life 2.* London: Sage.

Polkinghorne, D. (1997). Reporting qualitative research as practice. In W. Tierney & Y. Lincoln (Eds.), *Representation and the text* (pp. 3-21). Albany, NY: SUNY Press.

Punch, K. (1998). *Introduction to social research.* London: Sage.

Punch, M. (1994). Politics and ethics in qualitative research. In N. Denzin & Y. Lincoln (Eds.), *Handbook of qualitative research* (pp. 83-98). London: Sage.

Rath, J. (2001). Review essay. *Qualitative Research, 1*(1), 111-114.

Richardson, L. (1990). *Writing strategies.* London: Sage.

Richardson, L. (1991). Value constituting practices, rhetoric, and metaphor in sociology. *Current Perspectives in Social Theory, 11*, 1-15.

Richardson, L. (1992a). The consequences of poetic representation. In C. Ellis & M. Flaherty (Eds.), *Investigating subjectivity* (pp. 125-140). London: Sage.

Richardson, L. (1992b). The poetic representation of lives. *Studies in Symbolic Interaction, 13*, 19-27.

Richardson, L. (1993). Poetic, dramatics, and transgressive validity. *The Sociological Quarterly, 34*(4), 695-710.

Richardson, L. (1994). Writing. In N. Denzin & Y. Lincoln (Eds.), *Handbook of qualitative research* (pp. 516-529). London: Sage.

Richardson, L. (1996a). Educational birds. *Journal of Contemporary Ethnography, 25*(1), 6-15.

Richardson, L. (1996b). A sociology of responsibility. *Qualitative Sociology, 19*(4), 519-524.

Richardson, L. (1997). *Fields of play.* New Brunswick, NJ: Rutgers University Press.

Richardson, L. (1999). Feathers in our cap. *Journal of Contemporary Ethnography, 28*(6), 660-668.

Richardson, L. (2000). Writing. In N. Denzin & Y. Lincoln (Eds.), *Handbook of qualitative research* (2nd ed., pp. 923-948). London: Sage.

Richardson, L. (2001a). Getting personal. *Qualitative Studies in Education, 14*(1), 33-38.

Richardson, L. (2001b). Poetic representations of interviews. In J. Gubrium & J. Holstein (Eds.), *Handbook of interview research* (pp. 877-891). London: Sage.

Richardson, L., & Lockridge, E. (1991). The sea monster. *Symbolic Interaction, 14*, 335-341.

Richardson, L., & Lockridge, E. (1998). Fiction and ethnography. *Qualitative Inquiry, 4*(3), 328-336.

Richardson, M. (1998). Poetics in the field and on the page. *Qualitative Inquiry, 4*(4), 451-462.

Richardson, M. (1999). The anthro in cali. *Qualitative Inquiry, 5*(4), 563-565.

Riessman, C. (1993). *Narrative analysis.* London: Sage.

Rinehart, R. (1995). Pentecostal aquatics. *Studies in Symbolic Interaction, 19*, 109-121.

Rinehart, R. (1998a). Fictional methods in ethnography. *Qualitative Inquiry, 4*(2), 200-224.

Rinehart, R. (1998b). *Players all.* Bloomington, IN: Indiana University Press.

Rinehart, R. (2002). On "Sk8ing." In J. Denison & P. Markula (Eds.), *"Moving writing"* (pp. 145-158). New York: Peter Lang.

Ropers-Huilman, B. (1999). Witnessing. *Qualitative Studies in Education, 12*(1), 21-35.

Rorty, R. (1985). Solidarity or objectivity? In J. Rajchman & C. West (Eds.), *Post-analytic philosophy* (pp. 3-19). New York: Columbia University Press.

Rorty, R. (1989). *Contingency, irony, and solidarity.* Cambridge, UK: Cambridge University Press.

Rosaldo, R. (1993). *Culture and truth.* London: Routledge.

Rowe, D. (2000). *Amour impropre,* or "Fever Pitch" sans reflexivity. *Sociology of Sport Journal, 17*(1), 95-97.

Rowe, D. (2002). A fan's life. In J. Denison & P. Markula (Eds.), *"Moving writing"* (pp. 113-127). New York: Peter Lang.

Sandelowski, M. (1994). The proof of the pudding. In J. Morse (Ed.), *Critical issues in qualitative research methods* (pp. 46-63). London: Sage.

Sanders, C. (1995). Stranger than fiction. *Studies in Symbolic Interaction, 7,* 89-104.

Sandoz, J. (Ed.). (1997). *A whole other ball game.* New York: Noonday Press.

Sandoz, J., & Winans, J. (Eds.). (1999). *Whatever it takes.* New York: Farrar, Straus and Giroux.

Sarup, M. (1989). *An introductory guide to post-structuralism and postmodernism.* London: Harvester Wheatsheaf.

Schechter, J. (2001). The Indian subcontinent. *Qualitative Inquiry, 7*(1), 104-108.

Schwalbe, M. (1995). The responsibilities of sociological poets. *Qualitative Sociology, 18*(4), 493-414.

Schwandt, T. (1996). Farewell to criteriology. *Qualitative Inquiry, 2*(1), 58-72.

Schwandt, T. (1997). *Qualitative inquiry.* London: Sage.

Silvennoinen, M. (1993). A model for a man. In L. Laine (Ed.), *On the fringes of sport* (pp. 26-31). St. Augustin, Germany: Academia Verlag.

Silvennoinen, M. (1994a). A return to the past. *Young: Nordic Journal of Youth Research, 2*(4), 36-45.

Silvennoinen, M. (1994b). To childhood heroes. *International Review for the Sociology of Sport, 29*(1), 25-30.

Silvennoinen, M. (1999a). Anguish of the body. In A. Sparkes & M. Silvennoinen (Eds.), *Talking bodies* (pp. 93-98). Jyvaskyla, Finland: SoPhi.

Silvennoinen, M. (1999b). My body as metaphor. In A. Sparkes & M. Silvennoinen (Eds.), *Talking bodies* (pp. 163-175). Jyvaskyla, Finland: SoPhi.

Silvennoinen, M. (2002). Ecstasy on skis. In J. Denison & P. Markula (Eds.), *"Moving writing"* (pp. 159-170). New York: Peter Lang.

Simonelli, J. (2000). Field school in Chiapas. *Qualitative Inquiry, 6*(1), 104-106.

Sironen, E. (1994). On memory-work in the theory of body culture. *International Review for Sociology of Sport, 29*(1), 5-13.

Skelton, A. (2000). Review of *Talking bodies. European Physical Education Review, 6*(3), 27-29.

Smith, B. (1999). The abyss. *Qualitative Inquiry, 5*(2), 264-279.

Smith, J. (1988). The evaluator / researcher as person vs. the person as evaluator / researcher. *Educational Researcher, 17,* 18-23.

Smith, J. (1989). *The nature of social and educational inquiry.* Norwood, NJ: Ablex.

Smith, J. (1993). *After the demise of empiricism.* Norwood, NJ: Ablex.

Smith, J., & Deemer, D. (2000). The problem of criteria in the age of relativism. In N. Denzin & Y. Lincoln (Eds.), *Handbook of qualitative research* (2nd ed., pp. 877-896). London: Sage.

Smith, P. (1999). Food truck's party hat. *Qualitative Inquiry, 5*(2), 244-261.

Smith, S. (1993). Who's talking / who's talking back? *Signs, 18*(21), 392-407.

Smyth, J., & Shacklock, G. (1998). Behind the "cleansing" of socially critical research accounts. In J. Smyth & G. Shacklock (Eds.), *Being reflexive in critical educational and social research* (pp. 1-12). London: Falmer.

Sparkes, A. (1989). Paradigmatic confusions and the evasion of critical issues in naturalistic research. *Journal of Teaching in Physical Education, 8,* 131-151.

Sparkes, A. (1991). Toward understanding, dialogue, and polyvocality in the research community. *Journal of Teaching in Physical Education, 10*, 103-134.

Sparkes, A. (1992). The paradigms debate. In A. Sparkes (Ed.), *Research in physical education and sport* (pp. 9-60). London: Falmer Press.

Sparkes, A. (1994a). Life histories and the issue of voice. *Qualitative Studies in Education, 7*(2), 165-183.

Sparkes, A. (1994b). Self, silence and invisibility as a beginning teacher. *British Journal of Sociology of Education, 15*(1), 93-118.

Sparkes, A. (1995). Writing people. *Quest, 47*, 158-195.

Sparkes, A. (1996). The fatal flaw. *Qualitative Inquiry, 2*(4), 463-494.

Sparkes, A. (1997a). Ethnographic fiction and representing the absent Other. *Sport, Education and Society, 2*(1), 25-40.

Sparkes, A. (1997b). Reflections on the socially constructed physical self. In K. Fox (Ed.), *The physical self* (pp. 83-110). Champaign, IL: Human Kinetics.

Sparkes, A. (1998a). Athletic identity. *Qualitative Health Research, 8*(5), 644-664.

Sparkes, A. (1998b). Reciprocity in critical research. In J. Smyth & G. Shacklock (Eds.), *Being reflexive in critical educational and social research* (pp. 67-82). London: Falmer.

Sparkes, A. (1998c). Validity in qualitative inquiry and the problem of criteria. *The Sport Psychologist, 12*(4), 363-386.

Sparkes, A. (1999a). Exploring body narratives. *Sport, Education and Society, 4*(1), 17-30.

Sparkes, A. (1999b). The fragile body-self. In A. Sparkes & M. Silvennoinen (Eds.), *Talking bodies* (pp. 51-74). Jyvaskyla, Finland: SoPhi.

Sparkes, A. (2000). Autoethnographies and narratives of self. *Sociology of Sport Journal, 17*(1), 21-43.

Sparkes, A. (2001). Myth 94. *Qualitative Health Research, 11*(4), 538-552.

Sparkes, A. (2002). Fragmentary reflections of the narrated body-self. In J. Denison & P. Markula (Eds.), *"Moving writing"* (pp. 51-76). New York: Peter Lang.

Sparkes, A., & Silvennoinen, M. (Eds.). (1999). *Talking bodies.* Jyvaskyla, Finland: SoPhi.

Sparkes, A., & Smith, B. (1999). Disrupted selves and narrative reconstructions. In A. Sparkes & M. Silvennoinen (Eds.), *Talking bodies* (pp. 76-92). Jyvaskyla, Finland: SoPhi.

Sparkes, A., & Smith, B. (2002). Sport, spinal cord injury, embodied masculinities and the dilemmas of narrative identity. *Men & Masculinities, 4*(3), 258-285.

Squires, S., & Sparkes, A. (1996). Circles of silence. *Sport, Education and Society, 1*(1), 77-101.

Stacey, J. (1991). Can there be a feminist ethnography? In S.B. Gluck & D. Patai (Eds.), *Women's words* (pp. 111-119). London: Routledge.

St. Pierre, E. (1996). The responsibilities of readers. *Qualitative Sociology, 19*(4), 533-538.

Stanley, L. (1993). On auto/biography in sociology. *Sociology, 27*(1), 41-52.

Stanley, L., & Wise, S. (1993). *Breaking out again.* London: Routledge.

Strathern, M. (1987). Out of context. *Current Anthropology, 28*(3), 251-281.

Sudwell, M. (1999). The body bridge. In A. Sparkes & M. Silvennoinen (Eds.), *Talking bodies* (pp. 13-28). Jyvaskyla, Finland: SoPhi.

Sugden, J., & Tomlinson, A. (1999). Digging the dirt and staying clean. *International Review for the Sociology of Sport, 34*(4), 385-397.

Swan, P. (1998). Dreaming up, dreaming through and dreaming of people, places and paraphernalia. In C. Hickey, L. Fitzclarence, & R. Matthews (Eds.), *Where the boys are* (pp. 141-153). Geelong, Australia: Deakin University Press.

Swan, P. (1999). Three ages of changing. In A. Sparkes & M. Silvennoinen (Eds.), *Talking bodies* (pp. 37-47). Jyvaskyla, Finland: SoPhi.

Tedlock, D. (1987). Questions concerning dialogical anthropology. *Journal of Anthropological Research, 43*, 235-337.

References

Thorne, S. (1997). The art (and science) of critiquing qualitative research. In J. Mores (Ed.), *Completing a qualitative project* (pp. 117-132). London: Sage.

Tierney, W. (1993). The cedar closet. *International Journal of Qualitative Studies in Education, 6*(4), 303-314.

Tierney, W. (1997). Lost in translation. In W. Tierney & Y. Lincoln (Eds.), *Representation and the text* (pp. 23-36). Albany, NY: SUNY Press.

Tierney, W. (1998). Life history's history. *Qualitative Inquiry, 4*(1), 49-70.

Tierney, W. (1999). Guest editor's introduction. *Qualitative Inquiry, 5*(3), 307-312.

Tierney, W., & Lincoln, Y. (1997). Introduction. In W. Tierney & Y. Lincoln (Eds.), *Representation and the text* (pp. vii-xvi). Albany, NY: SUNY Press.

Tiihonen, A. (1994). Asthma. *International Review for the Sociology of Sport, 29*(1), 51-62.

Tiihonen, A. (2002). Body talks. In J. Denison & P. Markula (Eds.), *"Moving writing"* (pp. 77-85). New York: Peter Lang.

Tillmann-Healy, L. (1996). A secret life in a culture of thinness. In C. Ellis & A. Bochner (Eds.), *Composing ethnography* (pp. 76-108). London: Altamira Press.

Tinning, R. (1992). Teacher education pedagogy. In T. Williams, L. Almond, & A. Sparkes (Eds.), *Sport and physical activity* (pp. 23-40). London: E & FN Spon.

Tinning, R. (1998). "What position do you play?" In C. Hickey, L. Fitzclarence, & R. Matthews (Eds.), *Where the boys are* (pp. 109-120). Geelong, Australia: Deakin University Press.

Tinning, R. (2000). Book review of *Talking bodies. Sport, Education and Society, 5*(1), 89-101.

Travisano, R. (1998). On becoming Italian American. *Qualitative Inquiry, 4*(4), 540-563.

Travisano, R. (1999). Kansas City woman. *Qualitative Inquiry, 5*(2), 262-263.

Tsang, T. (2000). Let me tell you a story. *Sociology of Sport Journal, 17*(10), 44-59.

Van Maanen, J. (1988). *Tales of the field.* Chicago: University of Chicago Press.

Van Maanen, J. (1995a). An end to innocence. In J. Van Maanen (Ed.), *Representation in ethnography* (pp. 1-35). London: Sage.

Van Maanen, J. (Ed.). (1995b). *Representation in ethnography.* London: Sage.

Verma, N. (1991). *The crows of deliverance.* London: Readers International.

Wacquant, L. (1995). Review article. *Body & Society, 1*(1), 163-179.

Weems, M. (2000). Windows. *Qualitative Inquiry, 6*(1), 152-163.

Wellin, C. (1996). "Life at Lake Home." *Qualitative Sociology, 19*(4), 497-516.

Wendell, S. (1996). *The rejected body.* London: Routledge.

Wolcott, H. (1990). *Writing up qualitative research.* London: Sage.

Wolcott, H. (1994). *Transforming qualitative data.* London: Sage.

Wolcott, H. (1995). *The art of fieldwork.* London: Sage.

Wood, M. (2000). Disappearing. *Sociology of Sport Journal, 17*(1), 100-102.

Woods, P. (1999). *Successful writing for qualitative researchers.* London: Routledge.

Woolgar, S. (1988). *Science.* London: Tavistock.

Wright, J. (1995). A feminist post-structuralist methodology for the study of gender construction in physical education. *Journal of Teaching in Physical Education, 15,* 1-24.

Wright, J. (1996). Mapping the discourses in physical education. *Journal of Curriculum Studies, 28,* 331-351.

Wright, J. (2000). Bodies, meanings and movement. *Sport, Education, and Society, 5*(2), 35-49.

Young, K. (1991). Perspectives on embodiment. *Journal of Narrative and Life History, 1*(2 & 3), 213-243.

Young, K., White, P., & McTeer, W. (1994). Body talk. *Sociology of Sport Journal, 11,* 175-194.

Zeller, N., and Farmer, F. (1999). "Catchy, clever titles are not acceptable." *Qualitative Studies in Education, 12*(1), 3-19.

Index

A

Aaron, H. 51
Adams, N. 128
Agar, M. 9, 10, 154, 155, 156-157, 230
Anderson, L. 7-8, 192, 231, 226
Angrosino, M. 2, 152, 155, 158, 159, 165, 181, 182
Anzul, M. xi, 107
Athens, L. 48
Atkinson, P. x, xi, 2, 6, 8, 9, 12, 13, 15, 16, 23, 32, 33, 34, 61, 108, 113, 124, 126, 127, 149, 201, 229, 231, 233
Austin, D. 111, 112
autoethnographies
    active readership and 94-97
    assumptions about 91-94
    defined viii, 73-74
    examples 74-88
    reflections on 100-105
    as sacrament and witness 97-99
    self-indulgence and 88-91

B

Baff, S. 111, 113-114, 115
Bain, L. 7
Bakhtin, M. 92
Bandy, S. 79, 117
Banks, A. 6, 152, 159, 160, 183, 184, 201
Banks, S. 6, 152, 153, 159, 160, 183, 184, 201
Barone, T. 2, 52, 94, 95, 96, 155, 156, 159, 178, 179, 180, 181, 182
Beck, L. 40
Becker, H. xi
Bentham, J. 31
Bernstein, R. 200, 219
Best, J. 184
Bethanis, P. 174
Blumenfeld-Jones, D. 216-217, 223
Bochner, A. vii, 2, 14, 74, 89, 91, 94, 95, 96, 97, 102-103, 104, 128, 129, 130, 181, 186, 187, 218
Boman, J. 58, 70
Brackenridge, C. 65, 66, 71
Brady, I. 111
Brady, M. 182
Bridges, D. 40
Bronte, C. 207
Brown, L. 137, 138, 143
Bruce, T. 161
Bruner, J. 31, 205, 206

C

Carroll, L. 162
Ceglowski, D. 154, 155, 215, 217
Charmaz, K. 48, 89
Chekov, A. 207
Cherryholmes, C. 11
Cheyne, A. 32
Christensen, P. 174
Church, K. 89, 91-92
Cicero 30
Clarke, G. 19, 62
Clifford, J. 10, 21, 230
Clough, P. 111, 154
Coe, D. 17
Coffey, A. 2, 6, 8, 14, 15, 16, 17, 89, 108, 113, 124, 126, 127, 149, 179, 217-219, 229, 233
Cole, C. 7, 21, 44, 50
Coles, R. 181
confessional tales
    conventions of 59-61
    defined viii, 57-59
    reflections on 70-72
    in sports 61-70
Cooper, K. 68
crisis of representation 4-5, 9-12
criticisms of tales. See judging tales
Cushman, D. 44

D

Darden, A. 79, 117
Deemer, D. 219, 221, 222, 223, 224
Delamont, S. 2, 8, 6
Denison, J. 7, 79, 153, 161, 163, 174, 178, 179-180, 183, 187, 199, 200
Denzin, N. 2, 3, 4, 5-6, 7, 8, 24, 128, 129, 144, 155, 191, 192, 195, 196, 227
DeVault, M. 99, 192, 195, 223
Dévis, J. 56
Dewar, A. 19-20, 167
Diversi, M. 151, 152, 153, 156, 180
Donmoyer, J. 128, 130
Donmoyer, R. 128, 130
Dowling Naess, F. 119-123
Downing, M. xi, 107
Duncan, M.C. 79, 85-88, 100, 165, 166, 174, 179, 180
Duquin, M. 100

E

Eakin, P. 93
Eisner, E. x, 16, 33, 37, 155, 191, 192, 200, 204-205, 206-207, 219, 224, 226, 227-229, 231
Ellis, C. 2, 14, 74, 76, 89, 91, 94, 96, 97, 102, 103, 128, 129, 130, 186, 187, 188, 209-210, 213, 218
Ellsworth, E. 20
Ely, M. xi, 107, 113
epilogue ix, 225-234
ethnodrama
    audiences for 131-133
    authenticity of 133-137
    defined ix, 127-128
    example 137-144
    lived experience in 129-131
    problems 144-145
    reflections on 145-148
evaluation of tales. See judging tales
Evans, J. 7

F

Farmer, F. 30, 31
Faulkner, G. 56
Featherstone, M. 11
Fernandez-Balboa, J.-M. 7, 79
Feyerabend, P. 219
fictional representations
    "been there" claims 153-155
    creative fiction 157-160, 165-177
    creative nonfiction 155-157
    defined ix
    ethnographic fictions in sport 160-165
    potential of 180-183
    purpose behind 177-180
    reasons for using 149-153
    reflections on 187-189
    risks of 183-187
Fine, A. 55, 56
Fine, M. 18, 20, 21, 65, 71, 91
Finley, M. 154
Finley, S. 111, 154
Fischer, M. 4

Index

Fisher, W. 159
Fitch, M. 128
Flaherty, M. 2
Foley, D. 7, 199
Foucault, M. 81
Frank, A. 48, 97, 99, 102, 212-213
Frank, K. 151, 152, 153, 160, 178, 183
Freeman, M. 91
Fussell, S. 78, 79

G

Gamson, J. 19
Garratt, D. 193, 194, 220, 222, 224
Geertz, C. 10, 13, 22, 36, 178, 219, 220, 230
Gergen, K. 16, 22, 25, 34, 35, 36, 37, 92-93, 219
Gergen, M. 16, 22, 25
Glesne, C. 114, 115, 126
Golden-Biddle, K. 2, 29, 34, 47, 48, 49
Goldstein, T. 128, 129, 131, 133
Gore, J. 7, 19, 20
Gorelick, S. 22
Gottschalk, S. 188
Gould, D. 40, 42, 43, 44, 45, 46, 49, 50, 54
Grant, B. 7, 199
Gray, R. 90, 128, 132, 133-134, 136, 137, 144, 145, 146, 147, 148
Greenberg, M. 145
Guba, E. 28, 39, 193, 208
Gusfield, J. 31

H

Habermas, J. 37
Halas, J. 162-163, 180
Hall, A. 19, 62
Hammersley, M. 193, 194
Hanson, T. 51
Harvey, D. 11
Hastrup, K. 22
Hertz, R. 2, 17, 18, 90
Hickey, C. 7
history
    of qualitative research 3-9
    of scientific tales 30-31
Hodkinson, P. 193, 194, 220, 222, 224
Hollinger, R. 11
Holt, N. 56
Hones, D. 111
Hooks, B. 52
Huberman, M. 28, 39-40
Humberstone, B. 19, 62, 63, 71
Hume, D. 31

I

Innanen, M. 79, 103
Ivonoffski, V. 128, 132, 133-134, 136, 137, 144, 145, 147, 148

J

Jackson, D. 92, 116-117
Janesick, V. xi
Jennings, P. 51
Jevne, R. 58, 70
Jones, S. 111
Josselson, R. 192
Juana, D. 114, 115
judging tales
    different criteria for 199-201
    emerging criteria for 207-217
    questions about 191-192
    reflections on 219-224
    reviews of author's work 192-195
    self-indulgence charges 89-94
    transgression of boundaries and 195-198, 217-219

truth and validity issues 201-204
verisimilitude and 136, 204-207

K

Kaskisaari, M. 77
Kelly, P. 7
Kennelly, B. 124, 125
Kiesinger, C. 111, 112-113, 115
Klein, A. 53, 54, 61, 62, 161
Kluge, M.A. 68, 69
Kosonen, K. 76-77
Krane, V. 202
Kress, S. 188
Krieger, S. 91, 153
Krizek, R. 88, 151, 153-154

L

Ladson-Billings, G. 19
Laine, L. 76
Langer, S. 107
Lather, P. 10, 201, 202-203
Law, J. 33, 34
Lax, M. 213-214
Leder, D. 146
Lieblich, A. 192, 207-208
Lincoln, Y. x, 2, 3, 4, 5-6, 7, 8, 9, 11, 15, 21, 24, 39, 192, 207, 208, 214, 227
Livia, A. 182
Locke, K. 2, 29, 34, 47, 48, 49
Lockridge, E. 128, 192
Luke, C. 19
Lyon, D. 11
Lyons, K. 7, 21

M

Macdonald, D. 167
Manning, K. 208
Marcus, G. 4, 22, 44
Markula, P. 7, 79, 161, 187, 199, 200
McCall, M. 127, 128, 144
McCloskey, D. 12
McLaughlin, D. 179
McLeod, J. 94
McTaggart, R. 202
McTeer, W. 40
Megill, A. 12
Meola, T. 57
Mienczakowski, J. 128, 132-133, 135, 136, 147
Milchrist, P. 79
Miles, M. 28, 39-40
Miller, E. 79
Miller, M. 128, 130
Minh-ha, T. 90
Mitchell, R. 89
Morgan, D. 73
Morrison, B. 90
Mykhalovskiy, E. 89, 91, 92, 93, 94

N

Nelson, J. 12
Neumann, M. 155
Newman, M. 51
Nilges, L. 7, 11, 161, 162, 183, 187, 198

O

Olesen, V. 19, 23
Oliver, K. 7
Orner, M. 20

P

Paget, M. 128
Parrott, K. 79
Pedersen, K. 63-64
Pelias, R. 90, 185, 201, 210-211

Index

Picasso, P. 201
Pifer, D. 128, 129, 138
Plath, D. 10
Plato 30
Plummer, K. x, 2, 27, 29, 32, 33, 199
poetic representations
    advantages of 107-108
    criteria for judging 210-211
    criticisms of 194-195, 203
    defined viii-ix
    Louisa May's story 108-111
    reflections on 124-126
    in social sciences 111-116
    in sport 116-123
    transgressing boundaries with 195-198
Polkinghorne, D. 225
Porter, C. 191
Punch, K. 2-3
Punch, M. 71

Q

qualitative research
    history of 3-9
    realist tales 39-56
    reflections on x
    shifting landscape of viii, 1-25
quantitative research viii. *See also* scientific tales

R

Rath, J. 188, 222-223
realist tales
    defined viii, 1, 39-40
    experiential author(ity) of 41-44
    interpretive omnipotence and 46-51
    modifications of 51-54
    participant's point of view in 44-46
    reflections on 54-56
reflexivity and voice 16-24
Richardson, L. ix, 2, 10, 11, 12, 15, 17, 19, 22, 23, 24,
    25, 30, 31, 34, 35, 73, 97, 98, 103-104, 105, 108-
    111, 113, 115, 124, 128, 129, 154, 155, 183, 184,
    187, 192, 195, 197, 203, 208-209, 219, 224, 225,
    226, 227, 228, 232, 234
Richardson, M. 111-112
Riessman, C. 201
Rinehart, R. 7, 13, 79, 91, 153, 155, 161, 162, 174,
    183, 186, 205, 215, 226
Ropers-Huilman, B. 98
Rorty, R. 165, 219
Rosaldo, R. 230, 231
Rowe, D. 161, 174
Ruth, B. 51

S

Sandelowski, M. 33, 185, 186, 200, 207
Sanders, C. 71, 72, 183, 184, 226
Sandoz, J. 79, 117
Sarup, M. 11
Schechter, J. 112
Schwalbe, M. 194
Schwandt, T. 205-206, 220, 221, 223
scientific tales
    defined viii
    as dominant tale 27-28
    format for 28-29
    history of 30-31
    reflections on 36-37
    scientific style 31-36
Shacklock, G. 58, 59, 72
Silvennoinen, M. 78, 79, 100, 101, 116, 161
Simonelli, J. 112
Sinding, C. 128, 132, 133-134, 136, 137, 144, 145,
    147, 148

Sironen, E. 76
Skelton, A. 101, 102
Smith, B. 56, 74, 75, 112
Smith, J. 28, 34, 35, 39, 193, 199, 219, 220, 221, 222,
    223, 224
Smith, S. 21
Smyth, J. 58, 59, 72
Sparkes, A. 7, 28, 39, 40, 42, 52, 54, 56, 62, 63, 79,
    93, 96, 97, 98, 100, 102, 104, 116, 146, 161, 165,
    166, 167, 174, 179, 180, 187, 192, 193, 197, 198,
    199, 201, 204, 213, 214, 215, 217, 219, 223, 226
Sprat, T. 30
Squires, S. 56
St. Pierre, E. 195
Stacey, J. 22
Stanley, L. 19, 73, 96
Stegner, W. 207
Strathern, M. 14
Sudwell, M. 79, 101
Sugden, J. 66, 67
Swan, P. 79, 117-119

T

Tarulli, D. 32
Tedlock, D. 46
Thompson, H.S. 67
Thorne, S. 179
Tierney, W. x, 2, 10, 11, 24, 25, 160, 165, 179, 207,
    212, 227, 228, 232
Tiihonen, A. 77, 79
Tillman-Healy, L. 75-76
Tinning, R. 7, 79, 100, 101, 102, 167
Tomlinson, A. 66, 67, 68
Travisano, R. 112
Tsang, T. 7, 79, 80-85, 93-94, 95, 100
Tuval-Mashiach, R. 192
Twain, M. 207

U

Udry, E. 40

V

Van Maanen, J. 2, 8, 9, 10, 21, 40, 41, 44, 45, 46, 47,
    48, 50, 56, 57, 58, 59, 60-61, 71, 72, 157, 185, 186,
    212, 230
Verma, N. 232-233
Vinz, R. xi, 107

W

Wacquant, L. 146
Waugh, J. 187
Weems, M. 112
Weis, L. 18, 20
Wellin, C. 128, 131, 144
Wendell, S. 19
Weseen, S. 18
White, P. 40
Williams, R. 33, 34
Winans, J. 79, 117
Wise, S. 19
Wolcott, H. xi, 2, 9, 70, 115, 204, 230
Wong, L. 18
Wood, M. 174, 175, 178
Woods, P. xi, 2, 18, 125-126, 232
Woolgar, S. 36, 40
Wright, J. 7

Y

Young, K. 40, 43, 44, 45, 46, 49, 50, 54

Z

Zeller, N. 30, 31
Zilber, R. 192

About the Author

Andrew C. Sparkes, PhD, is a professor of social theory and the director of the Qualitative Research Unit in the School of Health & Sport Sciences at the University of Exeter in Exeter, Devon, England. He earned his doctor of philosophy and bachelor of education from Loughborough University, and his master of arts from Durham University. He has taught qualitative research methods to undergraduate and postgraduate students for 12 years.

Extensively published, Dr. Sparkes has developed an international reputation as an expert and innovative thinker in qualitative research circles. Sparkes is the editor of *Auto/Biography*, a refereed journal of the British Sociological Association Study Group. He serves on the advisory board for *Sport, Education & Society* and is an editorial board member for the *European Physical Education Review, Sociology of Sport On-Line, Agora,* and the *International Journal of Men's Health.*